Grigorij Jur'evic Grabovskij

Investigation of coherent microscopic defects inside the tunneling barrier of a Josephson junction

Experimental Condensed Matter Physics
Band 11

Herausgeber
Physikalisches Institut

Prof. Dr. Hilbert von Löhneysen
Prof. Dr. Alexey Ustinov
Prof. Dr. Georg Weiß
Prof. Dr. Wulf Wulfhekel

Eine Übersicht über alle bisher in dieser Schriftenreihe erschienenen Bände finden Sie am Ende des Buchs.

Investigation of coherent microscopic defects inside the tunneling barrier of a Josephson junction

by
Grigorij Jur'evic Grabovskij

KIT Scientific Publishing

Dissertation, Karlsruher Institut für Technologie (KIT)
Fakultät für Physik
Tag der mündlichen Prüfung: 21. Juni 2013
Referenten: Prof. Dr. Alexey V. Ustinov, Prof. Dr. Georg Weiss

Impressum

KIT Scientific Publishing

Karlsruher Institut für Technologie (KIT)
KIT Scientific Publishing
Straße am Forum 2
D-76131 Karlsruhe

KIT Scientific Publishing is a registered trademark of Karlsruhe Institute of Technology. Reprint using the book cover is not allowed.

www.ksp.kit.edu

Print on Demand 2014

ISSN 2191-9925
ISBN 978-3-7315-0210-4
DOI 10.5445/KSP/1000040666

Contents

1 Introduction

Quantum bits, or qubits, are a hot topic in physics at present. In fact, on a quantum computer some mathematical problems can be solved in much shorter time than on conventional computers. The key difference between a qubit and a classical bit lies in the laws of quantum mechanics. While the state of a bit is either ''0'' or ''1'', the state of a qubit is an arbitrary superposition $a|0\rangle + b|1\rangle$. In analogy to a two-state quantum systems, $|0\rangle$ and $|1\rangle$ are referred to as the ground and excited state, respectively, and a and b are complex numbers. Qubits must be controllable, i.e. one must be able to initialize them in an arbitrary state and perform a readout. It should be possible to couple qubits to establish quantum gates and to extend the qubit array to a desired number of qubits. Furthermore, qubits have to have very long coherence times to yield reliable results. These are the criteria postulated by D. DiVincenzo in the year 2000 which identify a quantum system as a qubit. A set of qubits fulfilling these requirements constitutes the base of a quantum computer.

Although a computer based on quantum manipulation has not yet been implemented, qubits also spark great interest in the scientific society due to the new possibility to study quantum phenomena. For example, qubits are excellent objects to gain a deeper understanding of decoherence effects that constitute, in fact, the link between quantum and classical physics. On the one hand, full control over one qubit requires a connection to the laboratory equipment. On the other hand, the qubit has to be sufficiently isolated from the noisy environment to ensure long coherence times. By coupling only two or three qubits it becomes possible to observe the complicated state evolution of the entangled systems by measuring each qubit independently. Furthermore, a many qubit system can be used as a so-called quantum simulator which may help to understand better some very peculiar quantum effects, such as high T_c superconductivity.

Many approaches exist how to realize qubits, for example, electron or nuclear spins, trapped ions or photons. Furthermore, one of the most promising ways is based on electrical circuits made of superconducting materials. This feature makes controlling and coupling very easy. Initialization, operation and readout of the qubit are achieved by using a current source, a voltmeter and a microwave generator. The interaction between qubits can be established either by inductive or capacitive coupling. Furthermore, superconducting qubits are made on wafers using standard lithography techniques.

Superconducting qubits are based on the Josephson effect. The Josephson junction (JJ) is a very thin (≈ 2 nm) insulating layer between two superconducting bulks, which allows a loss-free tunneling of the Cooper pairs. Due to the junction capacitance and the non-linear Josephson inductance, this type of qubits can be compared to non-linear resonators. Thus, they are multi-level systems with non-equidistant level separation, which is a crucial factor for their use as qubits. The two lowest states are defined as the ground state $|0\rangle$ and the excited state $|1\rangle$. There are three main types: charge, flux and phase qubits. Charge qubits work with only one Cooper pair that tunnels onto or off a superconducting island. The basis states of a flux qubit are superpositions of the current flowing clockwise or counterclockwise. A phase qubit is also based on a persistent current, where the states correspond to wavefunctions of the Josephson phase in an anharmonic potential. A few years ago, also combinations of these approaches were presented, e.g. those nick-named Quatronium, Transmon or Fluxonium. Recently, the ''3D-Transmon'' placed in a cavity reached coherence times which allow one to perform several thousand quantum operations, thus approaching the threshold of quantum error correction.

The insulating layer of the JJ is blamed to significantly contribute to the qubit decoherence. On the one hand, superconducting qubits suffer from the dielectric losses. On the other hand, the spectra of the qubits show avoided level crossings due to parasitic two-level systems coherently interacting with the qubit. The real nature of these systems is still under debate. The majority, however, was identified to be located inside the JJ. Among other theories, the most studied and promising model is that of atomic two-level tunneling systems, which are well known from amorphous solids like glasses. These sometimes very coherent quantum systems can be observed, in particular, in phase qubits due to the relatively large JJ. It was even suggested to use them as quantum memory or as auxiliary qubits.

Also for glasses, the two-level tunneling systems (TLSs) were found more by accident than design. Until the 70s, the Debye model successfully described the characteristics of solids. Defects were the main reason for phonon scattering at higher temperatures. However, at temperatures about 1 K and below, the wavelength of phonons is so long that single point-like impurities should not play a role. In fact, the mean free path of phonons is rather limited by the sample dimension. Therefore, it is all the more astonishing, that Zeller and Pohl found not only a large quantitative discrepancy but also a different temperature dependence of the thermal conductivity between crystalline and vitreous SiO_2. This was the starting point of a new field in solid state physics. Only one year after their publication, Phillips, and independently also Anderson *et al.*, developed the so-called tunneling model. They still blamed defects to be the reason for the different behavior. However, they assumed that some atoms, or a small group of atoms, have two stable configurations, which are coupled to each other via a small tunneling probability. Thus, tunneling systems are not only able to scatter, but also create and annihilate phonons. Obviously, the consequences are also an increased heat capacity and, in the case of charged tunneling entities, different dielectric characteristics. Within the 40 years of research, not only essentially all experiments are compatible with the tunneling model, but it also seems that the model is applicable to a large variety of amorphous materials which show a rather universal behavior at low temperatures.

Although in the past only expectation values of a large number of tunneling systems were studied, the tunneling model is based on a single tunneling system. The two stable states are represented by a double well potential the main parameters of which are the tunneling and the asymmetry energy. The coherence characteristics of a TLS are mainly determined by the tunneling probability. However, measurements and theory suggest that the variation of the asymmetry energy is responsible for the coupling of the tunneling system to its environment. Electric and elastic fields dictate which well is energetically more favorable. To better understand these two coupling types, we consider two examples. All tunneling systems in amorphous silicon are uncharged and couple therefore only to strain fields like phonons or deformation. In contrast, the tunneling systems in the salt KCl doped with Li^+-ions have large electric dipole moments and a negligible interaction with elastic fields.

The aim of this thesis was to prove whether the coherent quantum systems in the disordered AlO_x layer inside the JJ of a phase qubit behave as predicted by the tunneling model for the TLSs. While their electric dipole character is responsible for the coupling to the qubit, we investigated their interaction to an externally applied static strain.

In the second chapter, the two-state quantum systems are presented in general and the used simulation routine is described. The JJ and the mode of operation of the phase qubit are explained in the third chapter, while the tunneling systems are described in the fourth chapter. In the fifth chapter the experimental setup and results are presented showing excellent agreement with the tunneling model. From a chance observation of two coherently interacting TLSs inside the JJ, the theoretical description is derived using solely the assumptions of the tunneling model.

2 Theory of two-state systems

In this chapter the theory concerning the quantum mechanical behavior of a qubit is summarized. In the first part, the two-state system (TSS) is introduced. In particular, the Bloch sphere, used to represent the quantum states as vectors and to visualize the time evolution, is explained. By using the master equation in the Lindblad form decoherence effects are taken into account which is demonstrated on three standard protocols: Rabi oscillations, exponential decay and Ramsey fringes. In the second part, the simulation written by me during my diploma and PhD theses will be described in detail. It calculates the time evolution of a three level qubit coupled to two TSSs, later referred to as TLSs (two-level tunneling systems, Sec. 4).

2.1 A two-state system (TSS)

A many-state system is the basis of any coherent dynamical process on the quantum scale. Although a large number of physical realizations exists like spins, atoms or even macroscopic objects like a superconducting qubit, the time evolution of the state can be calculated by using the laws of quantum mechanics. As soon as the energetic difference between neighboring states is sufficiently large and the temperature of the heat bath to which it couples is below a critical value, the system can be approximated with a very high accuracy as a TSS consisting only of the ground and the first excited state. In this case, the system can be used as a qubit.

2.1.1 The Bloch sphere

The Hamiltonian for a TSS is written as

$$H = \frac{E}{2}\sigma_z + \frac{g_{\rm re}}{2}\sigma_x + \frac{g_{\rm im}}{2}\sigma_y = \frac{1}{2}\begin{pmatrix} E & g^* \\ g & -E \end{pmatrix}, \tag{2.1}$$

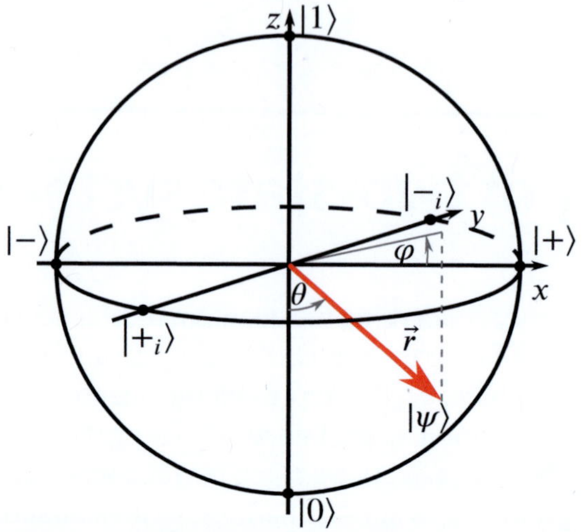

Figure 2.1: Bloch sphere. Quantum states and (x,y,z)-representations $|0\rangle \widehat{=} (0,0,-1)$, $|1\rangle \widehat{=} (0,0,1)$, $|\pm\rangle \widehat{=} (\pm 1,0,0)$, $|\pm_i\rangle \widehat{=} (0,\mp 1,0)$. The state $|\psi\rangle$ (red bold arrow) is defined by the angles θ and φ.

with σ_j ($j = \{x, y, z\}$) being the Pauli matrices [MD01] and the complex coupling strength between the two states $g = g_{\text{re}} + i\, g_{\text{im}}$. In the case of $g = 0$, H is diagonal and the eigenstates are the ground state $|0\rangle$ and the excited state $|1\rangle$ with the corresponding energies $\pm E/2$. The vectorial notation for theses states is

$$|0\rangle = \begin{pmatrix} 0 \\ 1 \end{pmatrix}, \; |1\rangle = \begin{pmatrix} 1 \\ 0 \end{pmatrix}. \tag{2.2}$$

An arbitrary state $|\psi\rangle$ of this system can be expressed as a superposition of the basis states with complex coefficients. These four degrees of freedom reduce to two after normalizing and eliminating the global phase, e.g. setting the phase of the ground state to zero. Thus, $|\psi\rangle$ can be expressed as

$$|\psi\rangle = \cos\frac{\theta}{2}|0\rangle + \sin\frac{\theta}{2}\, e^{-i\varphi}|1\rangle = \begin{pmatrix} \sin\frac{\theta}{2}\, e^{-i\varphi} \\ \cos\frac{\theta}{2} \end{pmatrix}. \tag{2.3}$$

This representation is very convenient because it illustrates the state graphically by a vector on the so called Bloch sphere (Fig. 2.1). The Bloch sphere is centered at the origin of a right-handed coordinate system with unit radius. The ground state $|0\rangle$ is placed on the bottom ($z = -1$) and the excited state $|1\rangle$

on top ($z = +1$). On the x-axis, the states $|\pm\rangle$ ($x = \pm 1$) and on the y-axis the states $|\pm_i\rangle$ ($y = \mp 1$) are located. These states are defined as follows:

$$|\pm\rangle = \frac{|0\rangle \pm |1\rangle}{\sqrt{2}}; \; |\pm_i\rangle = \frac{|0\rangle \pm i\,|1\rangle}{\sqrt{2}} \tag{2.4}$$

The definitions were chosen in a way that the prefactor in front of the ground state is always real and positive, and only the state $|1\rangle$ may have a complex phase. Although the cosine term in Eq. (2.3) might be negative for some θ, one can always make it positive by multiplying the state by -1. Using θ and φ as azimuthal and polar angles, respectively, the state ψ can be represented as a vector on the Bloch sphere. For example, $\theta = 0°$ corresponds to the ground state, and $\theta = \varphi = 90°$ results in the state $|-_i\rangle$. Since the Bloch sphere is not only spanned by the quantum states but also by the x-, y- and z-axis, it is useful to find also an expression of the quantum mechanical state in the (x, y, z)-representation. One can verify easily that the components of the position vector $\vec{r}$ can be written as

$$r_j = \langle\psi|\,\sigma_j\,|\psi\rangle\,,\; j = \{x, y, z\}, \tag{2.5}$$

and so

$$\vec{r} = \begin{pmatrix} \sin\theta\cos\varphi \\ \sin\theta\sin\varphi \\ -\cos\theta \end{pmatrix}, \tag{2.6}$$

which is known from the spherical coordinates.

Let us now come back to the Hamiltonian defined in Eq. (2.1). The eigenenergies are

$$E_{\pm} = \pm\frac{1}{2}\sqrt{E^2 + |g|^2}. \tag{2.7}$$

$|\psi_{\pm}\rangle$ denote the eigenstates. They can be expressed as

$$|\psi_-\rangle = \cos\frac{\alpha}{2}\,|0\rangle + \sin\frac{\alpha}{2}\,e^{-i\beta}\,|1\rangle = \begin{pmatrix} \sin\frac{\alpha}{2}\,e^{-i\beta} \\ \cos\frac{\alpha}{2} \end{pmatrix} \text{ and}$$

$$|\psi_+\rangle = \sin\frac{\alpha}{2}\,|0\rangle - \cos\frac{\alpha}{2}\,e^{-i\beta}\,|1\rangle = \begin{pmatrix} -\cos\frac{\alpha}{2}\,e^{-i\beta} \\ \sin\frac{\alpha}{2} \end{pmatrix}, \tag{2.8}$$

with

$$\tan\alpha = -\frac{|g|}{E}, \text{ and } g = |g|e^{i\beta}. \tag{2.9}$$

The eigenstates are orthogonal, since $\langle \psi_+ | \psi_- \rangle = 0$. Due to the factor 1/2, however, 180° have to be added to α of one eigenstate to get the orthogonal one. Thus, $|\psi_\pm\rangle$ always point into opposite directions on the Bloch sphere forming an axis, which will be very important when regarding the time evolution of the Bloch vector.

2.1.2 Time evolution and rotation

The time evolution of a two-state system starting with an arbitrary state $|\psi_0\rangle$ can easily be calculated analytically by solving the time dependent Schrödinger equation. In the following, the corresponding behavior on the Bloch sphere is shown. Let $|\psi_0\rangle$ have the form

$$|\psi_0\rangle = \cos\frac{\theta}{2}\,|0\rangle + \sin\frac{\theta}{2}\,e^{-i\varphi}\,|1\rangle = \begin{pmatrix} \sin\frac{\theta}{2}\,e^{-i\varphi} \\ \cos\frac{\theta}{2} \end{pmatrix}. \tag{2.10}$$

The time evolution of this state is given by the Schrödinger equation

$$\frac{d}{dt}\,|\psi(t)\rangle = -\frac{i}{\hbar}H\,|\psi(t)\rangle\,. \tag{2.11}$$

One can use the representation $|\psi(t)\rangle = U(t)\,|\psi_0\rangle$ with $U(t)$ being a unitary matrix. The differential equation is then solved by writing the eigenstates of H into the matrix B and the eigenvalues into the diagonal matrix[1] D, and so

$$U(t) = e^{-iHt/\hbar} = Be^{-iDt/\hbar}B^{-1} \tag{2.12}$$

$$= \begin{pmatrix} \cos\frac{\omega}{2}t + i\sin\frac{\omega}{2}t\,\cos\alpha & -i\sin\frac{\omega}{2}t\sin\alpha\, e^{-i\beta} \\ -i\sin\frac{\omega}{2}t\sin\alpha\, e^{i\beta} & \cos\frac{\omega}{2}t - i\sin\frac{\omega}{2}t\,\cos\alpha \end{pmatrix} \tag{2.13}$$

with

$$\omega = \frac{\sqrt{E^2 + g^2}}{\hbar}. \tag{2.14}$$

Comparing $U(t)$ with the standard expression for an unitary rotation matrix

$$R_{\vec{n}}(\delta) = \mathbb{1}\cos\frac{\delta}{2} - i\sin\frac{\delta}{2}\vec{n}\cdot\vec{\sigma} \tag{2.15}$$

[1] $H = BDB^{-1}$, $D = \begin{pmatrix} E_+ & 0 \\ 0 & E_- \end{pmatrix}$

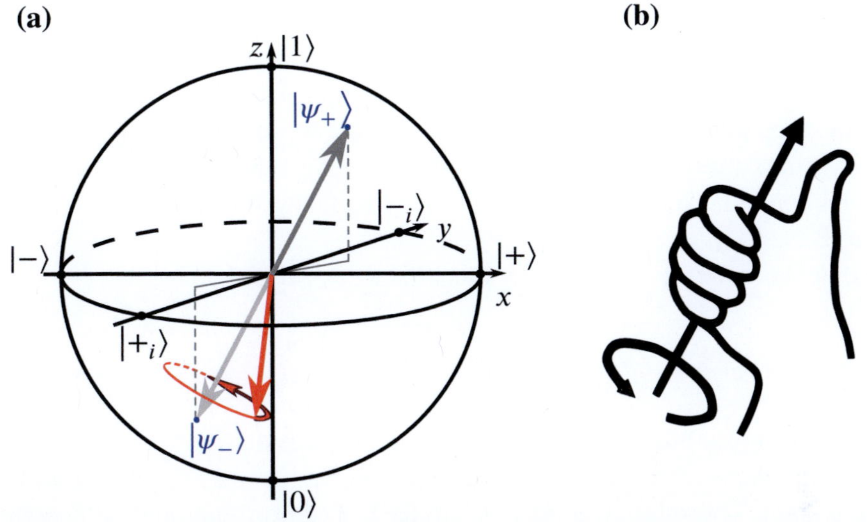

Figure 2.2: (**a**) Time evolution of a Bloch vector (red arrow): the vector rotates around the axis formed by the eigenstates of the Hamiltonian $|\psi_+\rangle$ (dark grey arrow) and $|\psi_-\rangle$ (grey arrow) with $E_+ > E_-$. The curved red arrow with the dark edge shows the direction of rotation. (**b**) Visualization of (a) with the right-hand rule. The thumb points into direction of the energetically higher state $|\psi_+\rangle$

which rotates a vector around the unit vector $\vec{n} = (n_x, n_y, n_z)$ by an angle δ [Lis08] ($\vec{\sigma} = (\sigma_x, \sigma_y, \sigma_z)$ symbolizes the set of Pauli matrices) yields $\delta = \omega t$ and $\vec{n}$ pointing exactly to the state $|\psi_+\rangle$ fulfilling Eq. (2.6)

$$\vec{n} = \begin{pmatrix} \sin\alpha\cos\beta \\ \sin\alpha\sin\beta \\ -\cos\alpha \end{pmatrix}. \tag{2.16}$$

So, for future experiments we can find the time evolution of the Bloch vector very easily. The matrix in Eq. (2.15), and thus also $U(t)$, are satisfying the right-hand rule. This means, if the thumb of the right hand is parallel to the rotation axis defined by the eigenstates and is pointing to the energetically higher state, then the other fingers show the direction of rotation of the Bloch vector (Fig. 2.2).

2.1.3 Density matrix and Lindblad equation

The formalism of the density matrix was introduced by Lev Landau [LL81] and independently by John von Neumann in 1927. The density matrix allows not only to describe a pure state, which is a superposition of the basis states, but also statistical mixtures. Therefore, this representation is crucial when considering the measurement of a quantum system coupled to a bath and so being subject to decoherence effects. The density matrix is able to describe temporal (many successive measurements on one system, e.g. one qubit) as well as spacial ensembles (one measurement involving many equal systems, e.g. spins in host crystals [Bus+11]).

The density matrix ρ describing a mixed state is defined as

$$\rho = \sum_j p_j \left|\psi_j\right\rangle \left\langle\psi_j\right|, \sum_j p_j = 1, p_j > 0, \tag{2.17}$$

where p_j is the probability to find the system in the pure state $\left|\psi_j\right\rangle$. The diagonal entries of the density matrix reflect the probabilities to measure a system in a particular state. Therefore, the trace has to be 1. The non-diagonal entries shed light on the phase coherence.

Let us consider some special cases to demonstrate these ideas. For the ground state and the excited state,

$$\rho_0 = |0\rangle\langle 0| = \begin{pmatrix} 0 & 0 \\ 0 & 1 \end{pmatrix}, \rho_1 = |1\rangle\langle 1| = \begin{pmatrix} 1 & 0 \\ 0 & 0 \end{pmatrix}. \tag{2.18}$$

For example, we have prepared the qubit in the excited state. After some time, the probability to measure the qubit in the state $|1\rangle$ will be only 0.5 due to relaxation effects. This is exactly the case of a statistical mixture without any coherence:

$$\rho_{\text{mixt}} = 0.5\rho_1 + 0.5\rho_0 = \begin{pmatrix} 0.5 & 0 \\ 0 & 0.5 \end{pmatrix}. \tag{2.19}$$

In contrast, the density matrix for the pure states $|\pm\rangle$ is

$$\rho_+ = |+\rangle\langle +| = 0.5\begin{pmatrix} 1 & 1 \\ 1 & 1 \end{pmatrix}, \rho_- = |-\rangle\langle -| = 0.5\begin{pmatrix} 1 & -1 \\ -1 & 1 \end{pmatrix}. \tag{2.20}$$

Here, the off-diagonal entries clearly identify these pure states. If the system in this case is not affected by relaxation but only by dephasing, only off-diagonal terms will vanish with time. Thus, both density matrices $\rho_\pm$ will pass into ρ_{mixt} (Eq. (2.19)). The former pure states turn into a statistical mixture where only the state population $P(|1\rangle) = \text{Tr}(\rho_{\text{mixt}}|1\rangle\langle 1|) = 0.5 = P(|0\rangle)$ is conserved. For a more detailed analysis and interpretation, please refer to literature [CDL06].

The Schrödinger equation in the density matrix formalism can be written as

$$\dot{\rho} = \sum_i p_i(|\dot{\psi_i}\rangle\langle\psi_i| + |\psi_i\rangle\langle\dot{\psi_i}|) = -\frac{i}{\hbar}[H,\rho]. \tag{2.21}$$

This expression describes the time evolution of an ensemble, which is, however, perfectly isolated from its environment. The simplest mathematical solution for also taking decoherence effects into account is the master equation in the Lindblad form [Lin76]

$$\dot{\rho} = -\frac{i}{\hbar}[H,\rho] + \sum_j \Gamma_j \mathscr{L}_j,$$

$$\mathscr{L}_j = L_j \rho L_j^\dagger - \frac{1}{2}(L_j^\dagger L_j \rho + \rho L_j^\dagger L_j). \tag{2.22}$$

L_j and Γ_j are the Lindbladian operator and the corresponding characteristic rate of the jth decoherence channel. There are three main decoherence channels which are briefly described in the following. We rewrite the 2x2 density matrix (Eq. (2.17)) in the general form

$$\rho = \begin{pmatrix} \rho_{11} & \rho_{10} \\ \rho_{10}^* & 1-\rho_{11} \end{pmatrix}. \tag{2.23}$$

The relaxation to the ground state is described by the characteristic decay time $T_1 = 1/\Gamma_1$. The corresponding matrix is the annihilation operator

$$\sigma_- = \begin{pmatrix} 0 & 0 \\ 1 & 0 \end{pmatrix}. \tag{2.24}$$

By setting $L_1 = \sigma_-$ we get

$$\mathscr{L}_1 = \begin{pmatrix} -\rho_{11} & -\rho_{10}/2 \\ -\rho_{10}^*/2 & \rho_{11} \end{pmatrix} \tag{2.25}$$

$\mathscr{L}_1$ has diagonal as well as off-diagonal nonzero entries and is therefore not only responsible for decay but also for dephasing. Usually, the creation operator $\sigma_+ = \sigma_-^\dagger$ is also taken into account. However, in our case it is negligible. The excitation frequency of the qubit is of the order of 6 GHz. In contrast, the temperature of 30 mK corresponds to $k_B T/h \approx 0.6$ GHz. The probability to have an excitation is given by the Boltzmann factor $e^{-hf/k_B T} \approx 5 \cdot 10^{-5}$.

The third decoherence channel is called pure dephasing. Setting the Lindblad operator L_2 to be equal $\sigma_z/\sqrt{2}$ results in[2]

$$\mathscr{L}_2 = \begin{pmatrix} 0 & -\rho_{10} \\ -\rho_{10}^* & 0 \end{pmatrix} \tag{2.26}$$

with only off-diagonal entries. The characteristic time $T_2^* = 1/\Gamma_2^*$ does not symbolize a complex conjugated time but it should be distinguished from the time T_2 (Sec. 2.1.6) defined as

$$\frac{1}{T_2} = \Gamma_2 = \frac{\Gamma_1}{2} + \Gamma_2^* = \frac{1}{2T_1} + \frac{1}{T_2^*}. \tag{2.27}$$

The complete Lindblad equation (Eq. (2.22)) for our case then is

$$\frac{d}{dt}\begin{pmatrix} \rho_{11} & \rho_{10} \\ \rho_{10}^* & 1-\rho_{11} \end{pmatrix} = -\frac{i}{\hbar}[H,\rho] + \begin{pmatrix} -\Gamma_1\rho_{11} & -\Gamma_2\rho_{10} \\ -\Gamma_2\rho_{10}^* & \Gamma_1\rho_{11} \end{pmatrix}. \tag{2.28}$$

In the case of large T_2^*, the dephasing time is limited by T_1 only so that $T_2 = 2T_1$. This is fulfilled by the ideal, fully isolated atom. The decoherence times T_1 and T_2 are usually extracted from the decay (Sec. 2.1.5) and a Ramsey measurement (Sec. 2.1.6) [Chi+03].

2.1.4 Rabi oscillations

Rabi oscillations are coherent oscillations between two states of a quantum system interacting with a resonant or near-resonant field. First observations were published by Isidor Rabi in 1938 (Nobel Price in 1944), who investigated the magnetic moments of atomic nuclei. Rabi oscillations appear for example when perturbing a spin 1/2-system with an oscillating magnetic field [Web10] or, in quantum optics, when an atom is passing a resonant cavity field [SZ97]. This rather complex effect can be simplified by using two standard approximations, which make it also possible to solve the master equation (2.22) analytically.

Rotating wave approximation (RWA) and Jaynes-Cummings-Hamiltonian (JCH)

To describe Rabi oscillations, E. Jaynes and F. Cummings used a quantized electro-magnetic field [JC63]. The main simplification is referred to as the

[2] σ_z symbolizes the noise of the energy and so of the Bloch vector oscillation frequency

rotating wave approximation (RWA). Here we follow the analysis of Zimmermann [Zim06]. The total Hamiltonian can be written as

$$H_{\mathrm{tot}} = H_{\mathrm{q}} + H_{\mu} + H_{\mathrm{int}}, \tag{2.29}$$

with H_{q} and $H_{\mu} = \hbar\omega_{\mu}(a^{\dagger}a + 1/2)$ being diagonal Hamiltonians of the qubit (Eq. (2.1)) and the microwave field, respectively. H_{int} is the interaction term which is usually expressed as

$$H_{\mathrm{int}} = \vec{E} \cdot \vec{d}. \tag{2.30}$$

$\vec{E}$ is the oscillating electrical field of the microwave and $\vec{d}$ is the dipole operator of the qubit. Using the field quantization, the electrical field is proportional to

$$\vec{E} \propto (a + a^{\dagger}). \tag{2.31}$$

Assuming the microwave field being polarized parallel to the x-axis, $\vec{E} = E_x \vec{e}_x$, it couples only to the x-component of the qubit dipole moment so that the dipole operator of the qubit can be simplified to

$$d_x \propto \sigma_x = \sigma_- + \sigma_+. \tag{2.32}$$

Obviously, polarization makes only sense for quantum systems like spins or atoms. However, the result can be fully applied to qubits in general using the advantages of the Bloch representation.

Thus, the interaction Hamiltonian can be written as

$$H_{\mathrm{int}} = \frac{\hbar\Omega_0}{2}(\sigma_- + \sigma_+)(a + a^{\dagger}) \tag{2.33}$$

with the coupling strength $\hbar\Omega_0$. In the Heisenberg picture, the corresponding time evolution of the operators a, $a^{\dagger}$ and $\sigma_{\pm}$ is

$$a(t) \propto e^{-i\omega_{\mu}t},\ a^{\dagger}(t) \propto e^{i\omega_{\mu}t} \text{ and} \tag{2.34}$$

$$\sigma_{\pm}(t) \propto e^{\pm i\omega_{\mathrm{q}}t}, \tag{2.35}$$

where $\omega_{\mu}/2\pi$ and $\omega_{\mathrm{q}}/2\pi$ are the microwave and qubit frequency, respectively. The fundamental idea of the RWA is to neglect the fast oscillating terms $\sigma_+ a^{\dagger}$ and $\sigma_- a$ [WL84], when $\omega_{\mu}/2\pi \approx \omega_{\mathrm{q}}/2\pi$. On long time scales, i.e. when the slow oscillating terms $\sigma_- a^\dagger$ and $\sigma_+ a$ show little variations, the integrated contribution of the fast oscillating terms averages very rapidly to zero since they rotate in opposite directions with angular frequencies of $\sim \pm 2\omega_\mu$ to each other. Moreover, both terms violate the conservation of energy since they imply a simultaneous excitation of the qubit and creation of a photon, or a decay of the qubit combined with an annihilation of a photon. This would be important only on a very short time scale. The resulting Hamiltonian is the Jaynes-Cummings-Hamiltonian (JCH). It has the form

$$H_{\mathrm{JCH}} = H_{\mathrm{q}} + H_{\mu} + \frac{\hbar\Omega_0}{2}(\sigma_- a^\dagger + \sigma_+ a) \tag{2.36}$$

Dressed states model

For further analysis we need a matrix representation of the JCH. In first instance, it is not obvious how to solve this problem since the JCH works with the electromagnetic field which has an infinite number of modes. However, the dressed states picture suggests a simple solution.

Assuming the microwave field to have n photons and the qubit to be in its ground state, the initial state is $|0, n\rangle$.

$$H_{\mathrm{JCH}} |0, n\rangle = \left(-\frac{E}{2} + \left(n + \frac{1}{2}\right)\hbar\omega_\mu\right)|0, n\rangle + \frac{\hbar\Omega_0}{2}\sqrt{n}\,|1, n-1\rangle \tag{2.37}$$

If, on the other hand, we consider the state $|1, n-1\rangle$ then

$$H_{\mathrm{JCH}} |1, n-1\rangle = \left(\frac{E}{2} + \left(n - \frac{1}{2}\right)\hbar\omega_\mu\right)|1, n-1\rangle + \frac{\hbar\Omega_0}{2}\sqrt{n}\,|0, n\rangle \tag{2.38}$$

We see that the JCH combines only the states $|0, n\rangle$ and $|1, n-1\rangle$. In other words, one excitation is oscillating between the qubit and the microwave field with a frequency scaling with $\sqrt{n}$. In the case of coherent field, the photon number has a Gaussian distribution yielding for the coupling strength $\Omega = \Omega_0\sqrt{\langle n\rangle}$. Since adding a multiple of the unit matrix does not affect the time evolution, we can subtract $n\hbar\omega_\mu \mathbb{1}_2$ from H_{JCH}, resulting in

$$H_{\mathrm{JCH}} = \frac{1}{2}\begin{pmatrix} E - \hbar\omega_\mu & \hbar\Omega \\ \hbar\Omega & \hbar\omega_\mu - E \end{pmatrix}. \tag{2.39}$$

The angular Rabi frequency Ω_R is defined as the difference between the eigenenergies $E_\pm$ (Eq. (2.7))

$$\Omega_R = \frac{E_+ - E_-}{\hbar} = \sqrt{\left(\frac{E}{\hbar} - \omega_\mu\right)^2 + \Omega^2}. \tag{2.40}$$

The matrix in Eq. (2.39) looks very similar to the Hamiltonian considered at the beginning of this chapter (Eq. (2.1)). The difference, on the one hand, is that the coupling between the states is a real number. This has no physical meaning since it follows from the assumption that the microwave field is polarized parallel to the x-axis. On the other hand, the energy of the excited state is lowered and the ground state energy increased by the energy of the microwave. In terms of the RWA, this means to go into the rotating frame and to express the phase evolution of the qubit states with respect to the field.

The formalism described in Sec. 2.1.2 allows to draw the time evolution of the qubit on the Bloch sphere. Since the coupling is real, the angle β in Eq. (2.16) is zero. If the qubit is in resonance with the microwave field, $E = \hbar\omega_\mu$, $E_+ > E_-$, $\alpha = 90°$ and so the Bloch vector rotates around the x-axis from the ground state to the excited state (Fig. 2.2) passing $|-_i\rangle$ and back to the ground state on the other side of the Bloch sphere. In the off-resonant limit, the rotation axis converges to the z-axis either with α going to zero ($E < \hbar\omega_\mu$) or approaching 180° ($E > \hbar\omega_\mu$). The corresponding frequency of Rabi oscillations $\Omega_R/2\pi$ will then be faster, but with a much smaller amplitude (Fig. 2.3a). It is necessary to mention here that if the microwave source is turned off and $\Omega = 0$, Eq. (2.39) still describes the state evolution of the qubit in the rotating frame defined by the microwave frequency $\omega_\mu/2\pi$ [WL84].

Damping of Rabi oscillations

After analyzing the driven qubit without any losses it is also important to consider decoherence effects. The coherent interaction between the qubit and the microwave field is a crucial factor for applications on longer time scales. To simplify matters, the microwave is assumed to be in resonance with the qubit. The Hamiltonian (Eq. (2.39)) can then be written as

$$H_{\mathrm{res}} = \frac{\hbar}{2}\begin{pmatrix} 0 & \Omega \\ \Omega & 0 \end{pmatrix}. \tag{2.41}$$

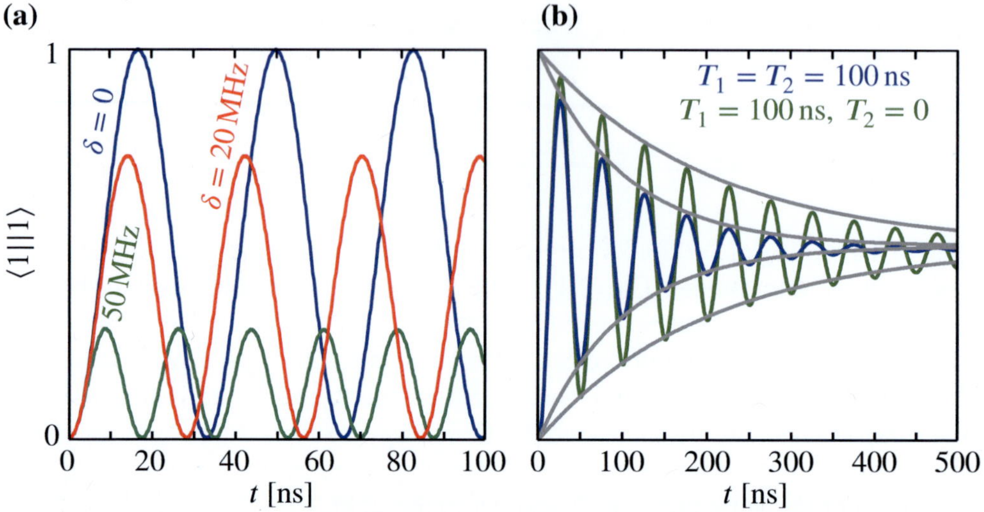

Figure 2.3: **(a)** Rabi oscillations without decoherence effects. δ is the detuning between microwave and qubit. **(b)** Two examples for damped Rabi oscillations are plotted

As discussed above, the eigenstates of H_{res} are $|\pm\rangle$ and so $\alpha = 90°$ and $\beta = 0°$. Starting from the ground state, the corresponding rotation matrix $U(t)$ (Eq. (2.13)) dictates the time evolution of the system

$$|\psi(t)\rangle = \begin{pmatrix} -i \sin \frac{\Omega}{2} t \\ \cos \frac{\Omega}{2} t \end{pmatrix}. \tag{2.42}$$

The Bloch vector always stays in the yz-plane. The density matrix for this undamped rotation is

$$\rho_{\text{no dec}}(t) = |\psi(t)\rangle \langle \psi(t)| = \frac{1}{2} \begin{pmatrix} 1 - \cos \Omega t & -i \sin \Omega t \\ i \sin \Omega t & 1 + \cos \Omega t \end{pmatrix}. \tag{2.43}$$

Rabi oscillations without decoherence are shown in Fig. 2.3(a).

In order to be able to compare this result with the density matrix taking decoherence effects into account (Eq. (2.43)), we define

$$\rho(t) = \begin{pmatrix} \rho_{11}(t) & -i\rho_{10}(t) \\ i\rho_{10}(t) & 1 - \rho_{11}(t) \end{pmatrix}. \tag{2.44}$$

with real functions ρ_{11} and ρ_{10}. Using the master equation in the form of Eq. (2.28) we find the differential equations for these two unknown functions that have to be solved:

$$\dot{\rho_{11}} = -\Gamma_1\rho_{11} + \Omega\rho_{10} \tag{2.45}$$

$$\dot{\rho_{10}} = -\Omega\rho_{11} - \Gamma_2\rho_{10} + \frac{\Omega}{2}. \tag{2.46}$$

An equivalent expression is

$$\begin{pmatrix} \dot{\rho_{11}} \\ \dot{\rho_{10}} \end{pmatrix} = \begin{pmatrix} -\Gamma_1 & \Omega \\ -\Omega & -\Gamma_2 \end{pmatrix} \begin{pmatrix} \rho_{11} \\ \rho_{10} \end{pmatrix} + \begin{pmatrix} 0 \\ \Omega/2 \end{pmatrix}. \tag{2.47}$$

One can easily verify that the solution of a differential equation of the form

$$\dot{y} = Ay + b \tag{2.48}$$

with a matrix A and a vector b is

$$y(t) = C \cdot e^{At} - A^{-1}b, \tag{2.49}$$

where C is the starting condition. We will see in further equations that the homogeneous part decays with time, yielding the steady state

$$\begin{pmatrix} \rho_{11} \\ \rho_{10} \end{pmatrix} \xrightarrow{t\to\infty} \frac{1}{2(\Gamma_1\Gamma_2 + \Omega^2)} \begin{pmatrix} \Omega^2 \\ \Omega\Gamma_1 \end{pmatrix}. \tag{2.50}$$

With typical decay rates of the qubit and the Rabi frequencies of

$$\Gamma_1 \approx \Gamma_2 \approx 10\text{MHz} < \Omega \approx 2\pi \cdot 30\text{MHz}, \tag{2.51}$$

Eq. (2.50) can be simplified to

$$\begin{pmatrix} \rho_{11} \\ \rho_{10} \end{pmatrix} \xrightarrow{t\to\infty} \begin{pmatrix} 0.5 \\ 0 \end{pmatrix}. \tag{2.52}$$

This is in agreement with the time independent part of the undamped oscillations in Eq. (2.43).

The procedure to find also the time dependent part of the solution is the same as in Eq. (2.12) by using eigenstates and eigenvalues. To simplify calculations, we apply the approximation

$$\left|\frac{\Gamma_1 - \Gamma_2}{\Omega}\right| \ll 1. \tag{2.53}$$

The eigenvalues of the matrix in Eq. (2.47) then are

$$\lambda_{\pm} = -\frac{\Gamma_1 + \Gamma_2}{2} \pm i\Omega, \tag{2.54}$$

and the eigenvectors simplify to

$$v_{\pm} = \begin{pmatrix} 1 \\ \pm i \end{pmatrix}. \tag{2.55}$$

After putting everything together, the final solution is

$$\begin{pmatrix} \rho_{11}(t) \\ \rho_{10}(t) \end{pmatrix} = \frac{1}{2} e^{-\frac{\Gamma_1+\Gamma_2}{2}} \begin{pmatrix} \cos\Omega t & \sin\Omega t \\ -\sin\Omega t & \cos\Omega t \end{pmatrix} \cdot C + \begin{pmatrix} 0.5 \\ 0 \end{pmatrix}. \tag{2.56}$$

By comparing this result to the density matrix in Eq. (2.43) we find the starting condition $C = (-1 \;\; 0)$ and so

$$\rho(t) = \frac{1}{2} e^{-t/\tau} \begin{pmatrix} -\cos\Omega t & -i\sin\Omega t \\ i\sin\Omega t & \cos\Omega t \end{pmatrix} + \begin{pmatrix} 0.5 & 0 \\ 0 & 0.5 \end{pmatrix}, \tag{2.57}$$

where τ is the characteristic time of the damping

$$\tau = \left(\frac{1}{2T_1} + \frac{1}{2T_2} \right)^{-1}. \tag{2.58}$$

Fig. 2.3(b) displays decaying Rabi oscillations simulated using the Lindblad equation (Sec. 2.2).

In our experiments, the qubit is excited by a so called π-pulse. This means that the qubit is performing Rabi oscillations for half a period. Accordingly, a $\pi/2$-pulse and a $3\pi/2$-pulse brings the Bloch vector of the qubit to the states $|-_i\rangle$ and $|+_i\rangle$, respectively.

2.1.5 Relaxation

Measuring the qubit relaxation directly provides T_1. After exciting the qubit by a π-pulse its excitation probability is recorded as a function of time. The Hamiltonian has the simple form

$$H = \frac{1}{2} \begin{pmatrix} E & 0 \\ 0 & -E \end{pmatrix}. \tag{2.59}$$

Since during relaxation no coherence is expected, the density matrix can be expressed as

$$\rho = \begin{pmatrix} \rho_{11} & 0 \\ 0 & 1-\rho_{11} \end{pmatrix} \tag{2.60}$$

with the initial condition $\rho_{11}(0) = 1$. Thus, the master equation (2.28) reduces to the differential equation

$$\dot{\rho_{11}} = -\Gamma_1 \rho_{11}, \tag{2.61}$$

and so

$$\rho_{11}(t) = Ce^{-\Gamma_1 t}. \tag{2.62}$$

with the decay rate Γ_1 and a constant C.

2.1.6 Ramsey fringes

Norman Ramsey was a PhD student of Isidor Rabi. Among other things, he explored the nuclear spin by using magnetic resonance techniques (Nobel Price in 1989). By measuring the Ramsey fringes, one is able to extract the dephasing time T_2, from which by knowing T_1 the pure dephasing time T_2^* can be calculated (Eq. (2.27)). The protocol for the pulse sequence is displayed in Fig. 2.4(a). The upper curve symbolizes the excitation spectrum of the qubit in dependence on the external flux (see next chapter). After the first $\pi/2$-pulse, the qubit evolves freely for some time t. During this period, it is useful to detune the qubit slightly from the microwave frequency, which is denoted by $\delta = \omega_\mathrm{q} - \omega_{\mu\mathrm{W}}$. Before the sequence is terminated by the readout, a second $\pi/2$-pulse is applied to the qubit. In terms of the RWA and the rotating frame picture we can rephrase this problem as follows: the first $\pi/2$-pulse brings the qubit to the state $|-_i\rangle$. The Bloch vector then starts to oscillates around the z-axis. Since after the time t another microwave pulse is applied, we only have to consider the accumulated phase of the Bloch vector in comparison to that of the microwaves, which is δt. Then, after the free evolution time, the qubit is mapped back on the state $|-_i\rangle$ because this state will be rotated to $|1\rangle$ by the second $\pi/2$-pulse. The decoherence effects during both pulses can be neglected, because on the one hand, the pulse duration is usually much shorter than T_1 and T_2, and on the other hand, the pulse duration is fixed yielding the same amount of decoherence for every measurement. Therefore, the relevant loss of information takes place between both pulses.

These ideas significantly simplify the calculations. The Hamiltonian for the free evolution of the Bloch vector in this case is

$$H = \frac{1}{2}\hbar \begin{pmatrix} \delta & 0 \\ 0 & -\delta \end{pmatrix}. \tag{2.63}$$

Using the general expression of the 2x2 density matrix

$$\rho = \begin{pmatrix} \rho_{11} & \rho_{10} \\ \rho_{10}^{*} & 1 - \rho_{11} \end{pmatrix}. \tag{2.64}$$

we can express the time dependence by solving the master Eq. (2.28) as

$$\rho(t) = \begin{pmatrix} C_0 e^{-\Gamma_1 t} & C_1 e^{-\Gamma_2 t} e^{-i\delta t} \\ C_1^{*} e^{-\Gamma_2 t} e^{i\delta t} & 1 - C_0 e^{-\Gamma_1 t} \end{pmatrix}, \tag{2.65}$$

where C_0 and C_1 are independent starting parameters. By setting

$$\rho(0) = |-_i\rangle\langle -_i| = \frac{1}{2}\begin{pmatrix} 1 & -i \\ i & 1 \end{pmatrix} \tag{2.66}$$

we find $C_0 = 1/2$ and $C_1 = -i/2$. Thus, the expectation value to measure the system in the state $|-_i\rangle$ or equivalently to find the system in the excited state after the whole pulse sequence is

$$\langle -_i|\rho(t)|-_i\rangle = \frac{e^{-\Gamma_2 t}}{2}\cos\delta t + \frac{1}{2}. \tag{2.67}$$

This expression describes oscillations with the detuning frequency $\delta/2\pi$ and the damping rate Γ_2. In Fig. 2.4(b), the blue curve with no detuning acts as the envelope for the red oscillating curve. A surface plot with varying detuning and time between the two $\pi/2$-pulses is displayed in Fig. 2.4(c). When performing the Ramsey experiment with only the time as parameter, a controlled detuning during the free evolution period is strictly recommended, because small detunings are hard to avoid in practice, and very slow oscillations affect the fitting procedure.

2.2 Simulation

To interpret and fit the experimental results and to be able to give theoretical predictions, it is useful to have a simulation of the quantum systems which uses

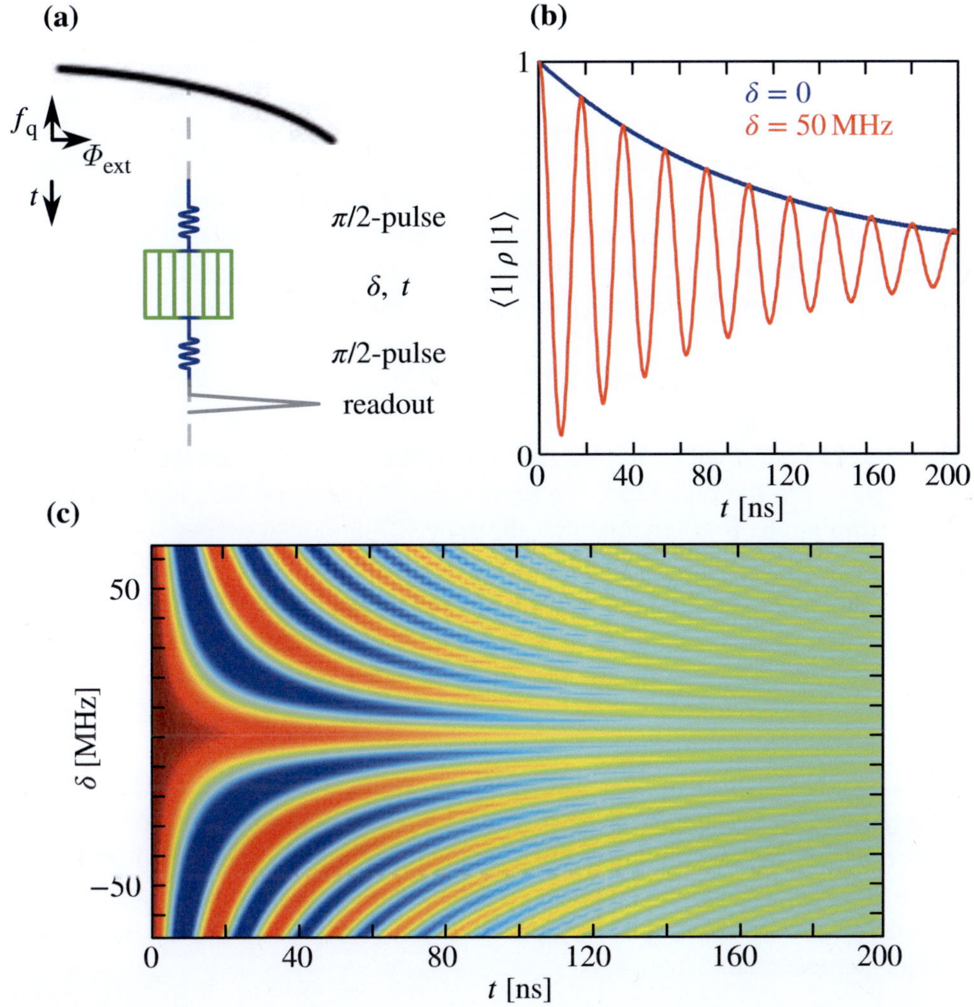

Figure 2.4: (**a**) Pulse sequence used in a Ramsey experiment. Upper part: resonance curve of the qubit as a function of the applied flux (see Sec. 3.3.3). Lower part: sequence of the Ramsey experiment (see text) (**b**) Exponentially decaying Ramsey fringes with no detuning (blue curve) and a detuning of 50 MHz (red curve). (**c**) A surface plot of the probability to measure the qubit in the excited state as a function of time and detuning (red maximal, blue minimal).

an algorithm mimicking the experimental procedure as well as possible. During my diploma thesis [Gra10], I implemented a simulation in the programming language Matlab®[3]. It took into account three coupled TSSs with decoherence by solving the Lindblad equation (2.22). Although this is a first-order differential equation, it is possible to calculate the time evolution of the quantum states exactly. The density matrix is therefore transformed into a vector of dimensionality $\dim(\rho)^2$ and a corresponding superoperator matrix is constructed. The time evolution of the density matrix can then be calculated via the usual procedure by using the eigenbasis and eigenvalues of the superoperator. All necessary steps are described in [CAG09]. The simulated results are in excellent agreement with measurements.

This simulation proved to be extremely helpful also during my PhD thesis. Therefore, I continued to implement new features as the knowledge about the TLSs grew. On the one hand, the second excited state of the qubit was added to the simulation. It is obvious that the state $|2\rangle$, and also higher states present in the phase qubit, may slightly influence the time evolution and measurement. In our case, however, the second excited state seems to play quite an active role as being one of the reasons for the shadow lines in the single photon spectra (Figs. 5.10 and 5.11). Furthermore, looking for a pulse sequence to measure the energy of the double excited state of two coupled TLSs, as described in Sec. 5.3.2, one possibility would be to use the $|2\rangle \leftrightarrow |1\rangle$ transition to populate the TLSs. On the other hand, simulating the experiments presented in this thesis strictly requires a generalization of the TLS Hamiltonians, in particular, they do not have to be diagonal initially (Sec. 4). To take the initial and projection states and the driving of the qubit by microwaves into account correctly, the simulation procedure had to be rewritten. In the following, the simulation routine is presented.

2.2.1 Simulation without decoherence

The idea of the simulation is to implement a control of the qubit energy and the microwave pulse sequence similar to the experiment. This is managed by creating a Hamiltonian for each step in the pulse sequence and letting the system evolve for the pulse duration according to this Hamiltonian (Fig. 2.5). Starting from, e.g. the ground state, the system undergoes several stepwise evolutions

[3] R2012a (7.14.0.739)

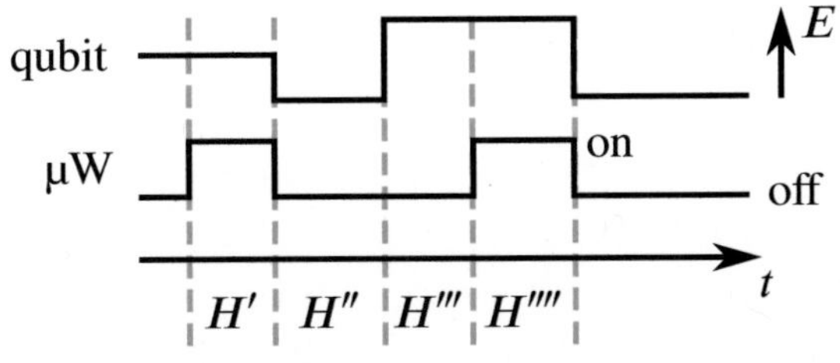

Figure 2.5: A pulse sequence in the experiment consists of a combination of varying qubit positions and switching the microwaves on and off. This can be modeled by the corresponding Hamiltonians H', H'' etc. and the corresponding time evolution matrices.

via the unitary matrix $U = \exp(-iHt/\hbar)$. The result is then projected on the state to be measured, e.g. the excited qubit.

The full Hamiltonian has the form

$$H = H_1 + H_{\mathrm{q}}^{(d)} + H_2 + H_{\mathrm{q}1} + H_{\mathrm{q}2} + H_{12}, \tag{2.68}$$

with the Hamiltonians of the respective systems H_1, $H_{\mathrm{q}}^{(d)}$, H_2 (TLS1, qubit, TLS2) and the corresponding coupling terms. For example, the TLS1 Hamiltonian is programmed as

$$H_1 = \left(\frac{1}{2}\varepsilon_1\sigma_z + \frac{1}{2}\Delta_1\sigma_x\right) \otimes \mathbb{1}_3 \otimes \mathbb{1}_2 = \left(\frac{1}{2}\varepsilon_1\sigma_{z,1} + \frac{1}{2}\Delta_1\sigma_{x,1}\right) \tag{2.69}$$

and its coupling to the qubit is

$$H_{\mathrm{q}1} = \frac{1}{2}\upsilon_1 \cdot \sigma_z \otimes \tau_x^{(d)} \otimes \mathbb{1}_2. \tag{2.70}$$

υ_1 is the coupling strength and σ_j the corresponding Pauli matrix. Since the qubit is simulated as a three state system and the dipole transition between the states $|2\rangle$ and $|1\rangle$ is $\sqrt{2}$-times larger than that between the first excited and the ground state (as it is the case in a harmonic oscillator [DCL06]),

$$\tau_x^{(d)} = \begin{pmatrix} 0 & \sqrt{2} & 0 \\ \sqrt{2} & 0 & 1 \\ 0 & 1 & 0 \end{pmatrix}. \tag{2.71}$$

The superscript (d) represents an operator acting on the eigenstates rather than physical states. Throughout this work, the phase qubit Hamiltonian is assumed to be already diagonal due to its similarity to the harmonic oscillator. This allows one to use $\tau_x^{(d)}$ as the coupling operator to electric fields, at least for the two lowest states. In the simulation, $H_{\mathrm{q}}^{(d)}$ is defined as

$$H_{\mathrm{q}}^{(d)} = \begin{pmatrix} E_{\mathrm{q}} - \Delta_{\mathrm{aq}} & 0 & 0 \\ 0 & 0 & 0 \\ 0 & 0 & -E_{\mathrm{q}}, \end{pmatrix} \tag{2.72}$$

with $\Delta_{\mathrm{aq}} = 2E_{|0\rangle\leftrightarrow|1\rangle} - E_{|0\rangle\leftrightarrow|2\rangle}$ being the asymmetry of the qubit.

The Hamiltonian defined in Eq. (2.68) can be used to find the spectrum of the three-partite system, and by fitting the experimental data, parameters like energies and coupling strengths can be found, and even different coupling types can be tested. This Hamiltonian also defines the unitary time evolution matrix. However, to be able to make time domain simulations, one has to think about the initial and projection states, in other words about the question of the basis. A very convenient basis consists of the eigenstates of individual uncoupled subsystems

$$\text{basis} = \{|e_1\rangle, |g_1\rangle\} \otimes \{|2\rangle, |1\rangle, |0\rangle\} \otimes \{|e_2\rangle, |g_2\rangle\}, \tag{2.73}$$

where $|g_j\rangle$ and $|e_j\rangle$ are the ground and excited state of the jth TLS, respectively. The problem is, however, how to transform H into this basis. This can be done in two steps. First, both TLS Hamiltonians are diagonalized resulting in

$$H^{(d)} = H_1^{(d)} + H_{\mathrm{q}}^{(d)} + H_2^{(d)} + H_{\mathrm{q1}}^{(d)} + H_{\mathrm{q2}}^{(d)} + H_{12}^{(d)}. \tag{2.74}$$

This step is straightforward because we go into the eigenbasis of the TLSs without destroying the order of subsystems in the Hamiltonian. The TLS1 Hamiltonian, for example, reads $H_1^{(d)} = E_1 \cdot (\sigma_z^{(d)} \otimes \mathbb{1}_3 \otimes \mathbb{1}_2)$ with $E_1 = \sqrt{\Delta_1^2 + \varepsilon_1^2}$. The coupling terms, of course, have to be transformed in the same way. This means, although all sub-Hamiltonians in $H^{(d)}$ are diagonal, now the coupling terms have all possible combinations of ''xx'', ''zx'', ''xz'' and ''zz''. From all these combinations, only the ''xx''-terms are responsible for energy transfer inducing anti-crossings and entangled states. The other coupling terms, on the other hand, are responsible for a change of the eigenstates (''xz'' and ''zx'') or just energy shifts (''zz''). Therefore, we can try to correct the change of eigenstates by going into the eigenbasis A of the matrix $H^{(d)} - H^{(d)}_{xx}$ with $H^{(d)}_{xx} = H^{(d)} * (\sigma_x^{(d)} \otimes \tau_x^{(d)} \otimes \mathbb{1}_2 + ...)$ (here $*$ means element wise multiplication)[4].

$$H^{(d,2)} = A^{-1} H^{(d)} A = A^{-1}(H^{(d)} - H^{(d)}_{xx}) A + A^{-1} H^{(d)}_{xx} A. \tag{2.75}$$

In contrast to the previous basis transformation, now the full 12x12 matrix is diagonalized so that the order of eigenenergies has to be reestablished by using the elements in A. Since the coupling strengths are much smaller than the TLS and qubit energies, it turns out that the new basis A is already almost Cartesian and so $A^{-1} H^{(d)}_{xx} A \approx H^{(d)}_{xx}$. The time evolution resulting from $H^{(d,2)}$ is still exact but referring to the basis which is practically the desired one, therefore it is used for simulation.

2.2.2 Driving and decoherence

A qubit driven by a microwave field undergoes Rabi oscillations which can be modeled in the RWA picture (Sec. 2.1.4). Therefore, we use the matrix

$$U_\mu(t) = \exp(-i\omega_\mu U_{\mu,0} t), \tag{2.76}$$

with the microwave frequency $\omega_\mu / 2\pi$ and

$$U_{\mu,0} = \frac{1}{2}(\sigma_z^{(d)} \otimes \mathbb{1}_3 \otimes \mathbb{1}_2 + \mathbb{1}_2 \otimes \tau_z^{(d)} \otimes \mathbb{1}_2 + \mathbb{1}_2 \otimes \mathbb{1}_3 \otimes \sigma_z^{(d)}). \tag{2.77}$$

By transforming $|\Psi\rangle$ into $|\Psi_\mu\rangle = U_\mu(t)\,|\Psi\rangle$, the time-dependent Schrödinger equation (2.11) can be rewritten with the corresponding Hamiltonian

$$H_\mu = -\frac{i}{\hbar}(U H^{(d,2)} U^\dagger + \hbar\omega_\mu U_{\mu,0}). \tag{2.78}$$

Depending on whether the microwave field is off or on, the Hamiltonian $H^{(d,2)}$ or the corresponding H_μ is used, which is referred to as H_{end}.

The Hamiltonian, and so the density matrix ρ, are 12x12 matrices. The idea of the Lindblad superoperator is to use the density matrix as a column vector ρ_v with 144 elements and to create a 144x144 matrix H_s fulfilling the relation

$$\dot{\rho}_v = H_s\, \rho_v. \tag{2.79}$$

[4]This procedure corresponds to the following trick. To find the eigenstates of the respective subsystems, the subsystems are far detuned from each other and then the Hamiltonian is diagonalized. In principle, this corresponds to simply neglecting the ''xx''-terms.

The time evolution of ρ_v can then be calculated analytically by finding the eigenstates and eigenenegies of H_s. As explained in [CAG09]

$$\begin{aligned} H_\mathrm{s} &= -i\left(H_\mathrm{end}\otimes \mathbb{1} - \mathbb{1}\otimes H^T_\mathrm{end}\right) \\ &+ \sum_j \Gamma_j \left(L_j\otimes L_j^* - \frac{1}{2}L_j^\dagger L_j \otimes \mathbb{1} + \frac{1}{2}\mathbb{1}\otimes \left(L_j^\dagger L_j\right)^T\right) \end{aligned} \tag{2.80}$$

with the Lindbladian operators L_j discussed in Sec. 2.1.3.

3 The phase qubit

As mentioned in the introduction, superconducting phase qubits can be compared with LC-resonators. They possess some capacitive and inductive elements, and therefore form a multi-level system. A simple LC-resonator is not suitable to serve as a qubit since due to the equidistant level separation addressing only the $|0\rangle \leftrightarrow |1\rangle$ transition is not possible. Trying to excite the $|1\rangle$-state would result in a Poissonian distribution of the population [Kul11]. One can circumvent this problem by ensuring a sufficiently large transition frequency difference between neighboring states, in particular between the states $|0\rangle$ and $|1\rangle$ and the states $|1\rangle$ and $|2\rangle$. This difference is referred to as the qubit asymmetry

$$\Delta_{\mathrm{aq}} = E_{|1\rangle\leftrightarrow|0\rangle} - E_{|2\rangle\leftrightarrow|1\rangle} = 2E_{|1\rangle\leftrightarrow|0\rangle} - E_{|2\rangle\leftrightarrow|0\rangle}. \tag{3.1}$$

To achieve $\Delta_{\mathrm{aq}} \neq 0$, a nonlinear element is required. In fact, all types of superconducting qubits are based on the Josephson junction (JJ), which will be introduced in Sec. 3.1. From all different superconducting qubit types, the phase qubit requires the largest JJ area to have a sufficiently large critical current. Since the two-level systems (TLSs) are located inside the JJ [Mar+05], it is best suited for investigating TLSs. The first proposal for a phase qubit was just a current biased JJ [DWM04]; [Mar09]. This approach was modified due to several reasons (Sec. 3.2.2). A better way is to enclose the JJ in a superconducting loop which is also know as the rf-SQUID [BP82]; [CB04]. Then, biasing is managed via an inductively coupled flux coil and for microwaves a waveguide connected to the JJ via a coupling capacitor is used. The principles of the rf-SQUID and the qubit operation will be discussed in Sec. 3.2, the realization and the experimental setup are described in Sec. 3.3.

3.1 The Josephson junction

3.1.1 The Josephson equations

Brian Josephson formulated his famous equations in 1974 [Jos74] and was honored with the Nobel Prize in 1973. He considered two superconductor bulks interrupted by a weak link. This is a small region across the superconducting line where the density of Cooper pairs is smaller than in the surrounding bulk. An example is a point-like contact of two superconductors. In our case, the weak link is a thin insulating layer forming a tunneling barrier between two overlapping superconducting lines. According to the Ginzburg-Landau theory [Sch+02], the superconducting state is described as a cooper pair condensate defined by a single wave function Ψ

$$\Psi = \Psi_0(\vec{x}, t)e^{i\phi(\vec{x},t)}, \tag{3.2}$$

where ϕ is a collective phase and the wave function is normalized to the density of Cooper pairs. When sending a super-current through the so-called Josephson junction (JJ), the Cooper pairs have to tunnel from one side of the weak link to the other resulting in a phase change across the JJ

$$\varphi = \phi_1 - \phi_2, \tag{3.3}$$

where ϕ_1 and ϕ_2 denote the phases of the two bulk superconductors. The Josephson equations depend only on the relative phase φ. The first Josephson equation predicts a super-current through the JJ

$$I_\mathrm{s} = I_\mathrm{c} \sin\varphi, \tag{3.4}$$

where I_c is the critical current or the maximal super-current. I_c depends on the parameters of the superconductor and on the geometry of the JJ. According to the first Josephson equation a current biased JJ with a current I_s less than I_c will still be superconducting and will not yield a voltage drop. The second Josephson equation becomes important when considering currents above I_c or alternating currents. Then, there is a nonzero time evolution of the Josephson phase resulting in a voltage drop V, which can be calculated from the second Josephson equation,

$$\dot{\varphi} = \frac{2e}{\hbar}V = \frac{2\pi}{\Phi_0}V \approx 2\pi \cdot 483.6\frac{\mathrm{MHz}}{\mu\mathrm{V}}\,V \tag{3.5}$$

where $\Phi_0 = h/2e$ is the magnetic flux quantum and e is the electron charge. Thus, the phase velocity $\dot{\varphi}$ is proportional to the voltage drop with the characteristic factor 483.6 MHz/µV.

3.1.2 The RCSJ model

The rather complex phase dynamics of a current biased JJ can be understood by considering the resistively and capacitively shunted JJ [Ste68]; [McC68]. The RCSJ model is only valid for small JJs with dimensions smaller than the spatial variations of the Josephson phase, typically about 30 µm. In contrast, in long JJs the spatial distribution of φ has to be taken into account as well resulting in much more complex differential equations [Wal01].

According to the RCSJ model, a small JJ can be modeled as three elements in parallel: a superconducting element according to Eqs. (3.4) and (3.5), an ohmic resistance R, which is responsible for damping and thus energy loss, and a capacitance C (Fig. 3.1(a)). According to Kirchhoff's law, the total current flowing through the junction yields the differential equation

$$0 = -I + I_c \sin\varphi + \frac{V}{R} + C\dot{V} = -I + I_c \sin\varphi + \frac{1}{R}\frac{\Phi_0}{2\pi}\dot{\varphi} + C\frac{\Phi_0}{2\pi}\ddot{\varphi}. \tag{3.6}$$

This equation can be compared to the equation of motion for a particle in a one-dimensional potential $U(x)$. According to classical Hamiltonian mechanics, the equation of motion reads

$$0 = m\ddot{x} + D\dot{x} + \frac{\partial U(x)}{\partial x}, \tag{3.7}$$

where m is the mass of the particle and D is the damping constant. By multiplying Eq. (3.6) with $\Phi_0/2\pi$ and identifying x with φ, we find the corresponding factors to be

$$m = C\left(\frac{\Phi_0}{2\pi}\right)^2, \quad D = \frac{1}{R}\left(\frac{\Phi_0}{2\pi}\right)^2, \tag{3.8}$$

and

$$U_J(\varphi) = \frac{I_c\Phi_0}{2\pi}\left(-\frac{I}{I_c}\varphi - \cos\varphi\right). \tag{3.9}$$

By introducing the Josephson energy $E_\mathrm{J} = I_\mathrm{c}\Phi_0/2\pi$ and normalizing the total current flowing through the junction $\gamma = I/I_\mathrm{c}$, Eq. (3.9) can be rewritten as

$$U_\mathrm{J}(\varphi) = E_\mathrm{J}(-\gamma\varphi - \cos\varphi). \tag{3.10}$$

The resulting washboard potential is plotted in Fig. 3.1(b) for three different values of γ. For currents larger than I_c, the potential barrier disappears and the particle starts to run down the potential which results in a voltage drop across the JJ. Due to quantum tunneling, however, the bias current does not have to overcome I_c to cause a voltage drop, but, depending on the JJ parameters and the temperature, currents very close to the critical current might be sufficient, yielding other interesting effects like phase diffusion [MK89]. The angular resonance frequency ω_0 can be obtained by solving the differential Eq. (3.7). For small underdamped JJs [1] the expression for ω_0 reads

$$\omega_0 = \sqrt{\frac{1}{m}\frac{\partial^2 U}{\partial\varphi^2}} = \omega_\mathrm{p}\sqrt{\cos\varphi} = \omega_\mathrm{p}\sqrt[4]{1-\gamma^2}, \tag{3.11}$$

with the angular plasma frequency

$$\omega_\mathrm{p} = \sqrt{\frac{2\pi I_\mathrm{c}}{\Phi_0 C}}. \tag{3.12}$$

One can see from Eq. (3.11) that the resonance frequency decreases with increasing bias current. At the bias current $I = I_\mathrm{c}$, ω_0 is zero and the potential well disappears. Here we introduce the Josephson inductance L_J according to the relation

$$\omega_0 = \frac{1}{\sqrt{L_\mathrm{J}C}}. \tag{3.13}$$

Thus,

$$L_\mathrm{J} = \frac{\Phi_0}{2\pi}\frac{1}{I_\mathrm{c}\cos\varphi} = L_{\mathrm{J},0}\frac{1}{\cos\varphi}, \tag{3.14}$$

where we have defined $L_{\mathrm{J},0} = \Phi_0/2\pi I_\mathrm{c}$. Note that the expression for L_J is consistent with the standard definition of the inductance $L = V/\dot{I}$. However, the nature of this inductance is not a stored energy in the magnetic field.

[1] Small damping factor corresponds to large R and so a very small current through the resistance.

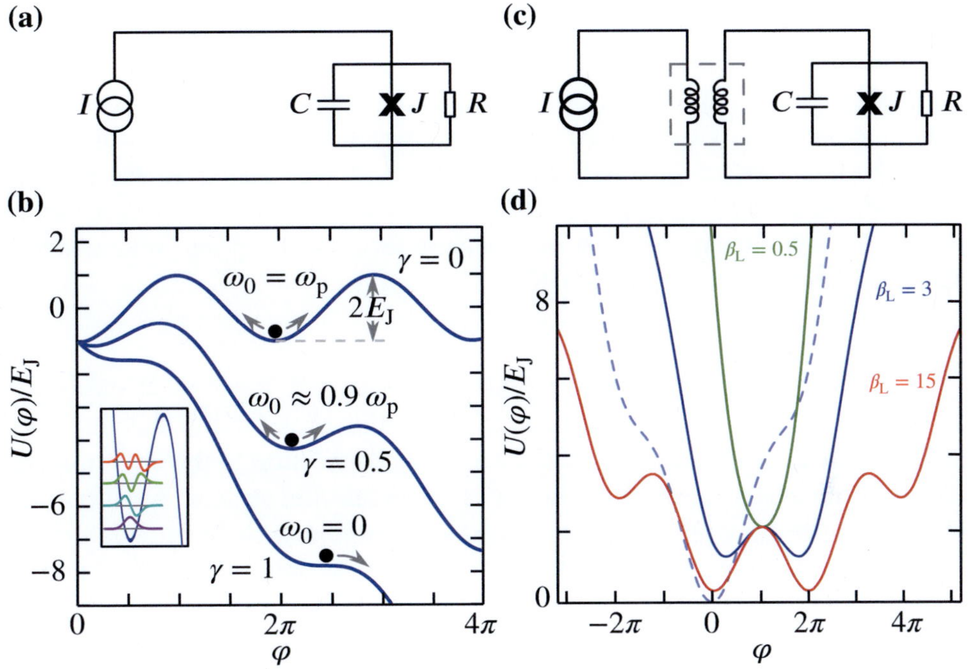

Figure 3.1: (a) The circuit diagram for a current biased JJ according to the RCSJ model is sketched and the resulting washboard potential is drawn in **(b)** for different bias currents. The inset shows the quantization of a potential well. **(c)** The circuit diagram for a flux biased rf-SQUID is shown schematically. **(d)** The potential corresponding to (c) is plotted for different β_L parameters at $\Phi_{ext} = \Phi_0/2$. The potential for $\beta_L = 3$ is also shown for $\Phi_{ext} = 0$ (dashed, light blue curve)

The inductance can be rather explained by the kinetic inductance [DWM04]; [Sch+02] known, e.g., from the behavior of superconductors at high frequencies. Most strikingly, L_J can be also negative and it can take any large number.

3.2 The phase qubit – theory

3.2.1 The current biased JJ as a phase qubit

In fact, first phase qubits were initially designed from current biased Josephson junctions [DWM04]; [Mar09]. The plasma frequency ω_p is fixed by design parameters. By current biasing the JJ during the experiment, the resonance frequency ω_0 can be adjusted (Eq. (3.11)). Then, the qubit operation is performed by combining the dc-current with the microwave signal. The readout is managed by a short current pulse. This tilts the potential for a short time so that the particle, if being in the excited state, tunnels out from its well and starts to run down the potential (Fig. 3.1(b)). Thus, by measuring the voltage drop across the JJ the final state of the qubit can be determined. The first experimental data were published by Martinis *et al.* in 2002 [Mar+02] which proved that a current biased JJ can perform all operations needed for a single qubit: initialization, Rabi oscillations and readout [DiV00].

3.2.2 The rf-SQUID

To be able to increase the coherence times, the phase qubit had to be re-designed. A current biased JJ requires a direct connection to the environment, which introduces a significant amount of noise into the system. Furthermore, the impedance of a JJ, $Z_\mathrm{J} = \sqrt{L_\mathrm{J}/C}$, is comparable with the standard $50\,\Omega$ impedance opening a significant decay channel [Lis08]. The solution is to enclose the JJ in a superconducting loop [Mar09] thus forming an rf-SQUID (Fig. 3.1(c)). The loop is galvanically disconnected from the environment, but it is still inductively coupled to the lines via a transformer, which significantly modifies the impedance of the JJ circuit. The readout is managed by a dc-SQUID magnetometer [Sch+02] coupled inductively to the qubit (see next section).

Fig. 3.1(c) illustrates the rf-SQUID used as the phase qubit which is inductively coupled to the current source. A detailed analysis of this circuit can be found in [BP82]. The externally applied magnetic flux Φ_ext induces a current in the loop which adds to the external flux. The resulting flux Φ_tot through the rf-SQUID loop can be expressed as

$$\Phi_\mathrm{tot} = \Phi_\mathrm{ext} + L_\mathrm{geo} I_\mathrm{q} \tag{3.15}$$

where L_geo is the geometrical inductance of the loop and I_q the circulating current. Due to flux quantization in a superconducting loop with the phase φ across the JJ as a degree of freedom, we find φ to be [Sch+02]

$$\varphi = 2\pi \frac{\Phi_\mathrm{tot}}{\Phi_0}. \tag{3.16}$$

Thus, the circulating current I_q can be expressed as

$$I_\mathrm{q} = \frac{\Phi_\mathrm{tot} - \Phi_\mathrm{ext}}{L_\mathrm{geo}} = \frac{\Phi_0}{2\pi L_\mathrm{geo}} \left(\varphi - 2\pi \frac{\Phi_\mathrm{ext}}{\Phi_0} \right). \tag{3.17}$$

The rf-SQUID potential $U(\varphi)$ is the sum of the junction energy $U_J(\varphi)$ given in Eq. (3.10) [Lis08] and the magnetic energy $L_{geo}I_q^2/2$.

$$U(\varphi) = E_J\left(1 - \cos\varphi + \frac{(\varphi - 2\pi\Phi_{ext}/\Phi_0)^2}{2\beta_L}\right), \tag{3.18}$$

where β_L is defined as

$$\beta_L = \frac{2\pi L_{geo}I_c}{\Phi_0} = \frac{L_{geo}}{L_{J,0}}, \tag{3.19}$$

the ratio of the geometric and the Josephson inductance at $\varphi = 0$. The rf-SQUID potential is plotted for different β_L values in Fig. 3.1(d). Aiming to operate the rf-SQUID as a phase qubit (Sec. 3.3.3), it is recommended for β_L to satisfy

$$1 < \beta_L < 4.6. \tag{3.20}$$

There are two reasons for this condition. On the one hand, it should be possible to initialize the qubit. Therefore, at a particular flux bias, e.g. $\Phi_{ext} = 0$, the potential $U(\varphi)$ should have only a single global minimum facilitating the initialization of φ always to the same value. This defines the upper limit of β_L. On the other hand, the readout of the qubit is only possible if there are at least two wells in the qubit potential. By applying a flux pulse, the potential is tilted so that the particle tunnels to the other well if it was in the excited state. This defines the lower condition for β_L.

3.2.3 The dc-SQUID

In contrast to the rf-SQUID, which has only one junction in its loop and is driven via ac-current, the dc-SQUID uses two junctions connected in parallel (Fig. 3.2). As the name suggests, a dc-current is flowing through the two branches [Sch+02]. By referring to the two branches as *a* and *b*, and assuming both junction to have the same critical current I_c, the current sent through the dc-SQUID is

$$I = I_a + I_b = I_c(\sin\varphi_a + \sin\varphi_b) = 2I_c\cos\frac{\varphi_a - \varphi_b}{2}\sin\frac{\varphi_a + \varphi_b}{2}. \tag{3.21}$$

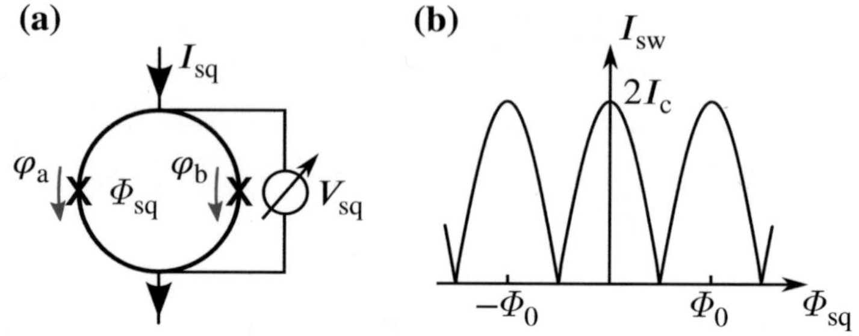

Figure 3.2: (a) The dc-SQUID is formed by two JJs enclosed in a superconducting loop. The dc-current I_{sq} is sent through both JJs connected in parallel while measuring a voltage drop V_{sq} across the SQUID. **(b)** The switching current of the SQUID, I_{sw}, depends periodically on the flux through the SQUID loop.

Due to the two phases of the junctions, a dc-SQUID has two degrees of freedom. However, by using the relation of the flux quantization, one degree can be eliminated:

$$\varphi_a - \varphi_b = 2\pi \frac{\Phi_{sq}}{\Phi_0}, \tag{3.22}$$

and so

$$I = 2I_c \cos\left(\pi \frac{\Phi_{sq}}{\Phi_0}\right) \sin\left(\varphi_b + \pi \frac{\Phi_{sq}}{\Phi_0}\right). \tag{3.23}$$

Φ_{sq} is the flux inside the SQUID loop, which can be controlled by an external coil. As can be seen from Eq. (3.23), the maximal super-current which can flow through the device strongly depends on the external flux. The equation has only one degree of freedom, which is inside the sine-term. Therefore, we can substitute all other parts of Eq. (3.23) by a prefactor corresponding to the maximal, or the critical, current of the device

$$I_{c,sq} = 2I_c \left|\pi \frac{\Phi_{sq}}{\Phi_0}\right|. \tag{3.24}$$

In the definition of the critical current of the SQUID, the absolute values are used. Obviously, a negative maximal current makes no sense, and furthermore, a negative sign can be taken into account in the sine-term by adjusting φ_b.

Being inductively coupled to a phase qubit, the switching current of the dc-SQUID reveals the flux produced by the qubit loop and so its state. Varying the qubit flux bias in a wide range, one can see the qubit steps due to tunneling of the qubit virtual particle between different wells [Lis08].

3.3 The phase qubit – experimental

3.3.1 The samples

For this thesis, two different qubit chips were used. Both chips were kindly provided to us by the UCSB group (Prof. John Martinis). Sample 1 represents the first generation of flux controlled phase qubits, whereas in sample 2 some important improvements were realized increasing the coherence by a factor of ten. Consequently, it has much less strongly coupled TLSs. However, it allows using time-domain measurements which are much more sensitive in retrieving TLS resonances.

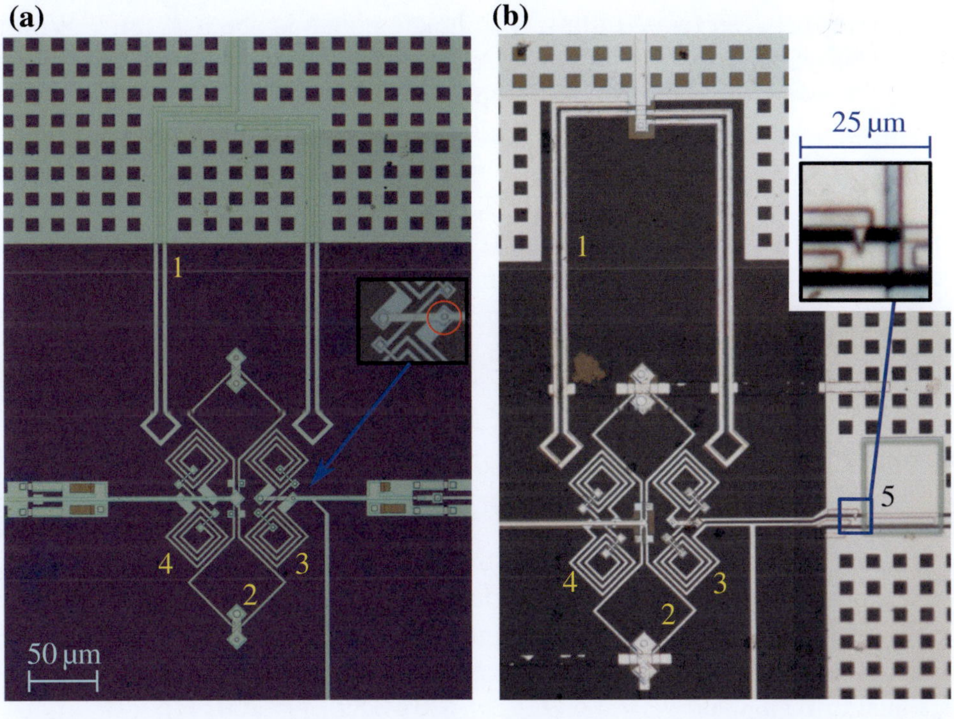

Figure 3.3: The two chips used in this thesis were provided to us by the group of J. Martinis (UCSB). The design of the flux bias lines (1) is gradiometric for the SQUID (2), but not for the qubit (3) which allows to control the qubit's frequency. On the other side of the SQUID an auxiliary structure is placed (4) for reasons of symmetry.

Sample 1 [Sim+04]

A photo of the chip is shown in Fig. 3.3(a). The group of Martinis attaches great importance to the gradiometric design. The flux bias lines and the dc-SQUID are arranged in such a way that the integral over the flux in the SQUID loop is zero. In contrast, the qubit is placed closer to one bias line than to the other resulting in a non-zero flux bias. On the opposite side of the SQUID an auxiliary structure is added for symmetry reasons. Thus, it is ensured that the SQUID magnetometer is always at the best working point resulting in a readout of the qubit with high fidelity. Furthermore, the lower part of all three loops is twisted by 180° with respect to the corresponding upper part to annihilate the magnetic noise due to homogeneous field fluctuations.

Silicon is used as the substrate. The tunneling barrier of the JJ is a disordered AlO_x layer between two Al films. The junction area is about 32 μm^2. With a critical current density of 40 A/cm^2, I_c equals 12 μA [Sim+04]. The ratio of the loop inductance of 168 pH to the Josephson inductance ($L_{J,0}$ = 28 pH) results in a β_L parameter (Eq. (3.19)) of 6. This value is too large to be used as a phase qubit without additional auxiliary means. The critical current of an identically fabricated sample is given in [Joh+05] as 8.5 μA and so β_L = 4.5. Assuming the same critical current density, the JJ area is 21 μm^2. The self capacitance of the JJ equals 1.2 pF. The relaxation time T_1 was measured by Simmonds *et al.* to be 41 ns. Our sample, which has been designed and produced in the same way, possesses a T_1 time of only about 13 ns, 9 years after the fabrication.

Sample 2 [Ste+06]

Sample 2 is shown in Fig. 3.3(b). The layout is basically unchanged in comparison to sample 1. However, there are several critical improvements. Now, sapphire is used as the substrate due to its very low loss tangent [Kru+06]. To reduce the number of anti-crossigns in the spectrum, the JJ area is scaled down to 1 μm^2. The critical current density is increased to 170 A/cm^2 yielding I_c equal to 1.7 μA and β_L = 3.6 (L_{geo} = 720 pH, $L_{J,0}$ = 200 pH).

The self capacitance of the JJ equals C_J = 50 fF. To reduce the resulting plasma frequency $\omega_p/2\pi$ (Eq. (3.12)), the JJ is shunted by a large plate capacitor with an area of 60×60 μm^2 and silicon-nitride as the dielectric [Mar+05]. This capacitance, C_{ext}, is designed to be 800 fF, yielding $C = C_J + C_{ext}$ = 850 fF. The decoherence times T_1 and T_2 were measured to be 110 and 105 ns, respectively.

Although we have been taking data with this qubit for 7 years already, the decoherence times have not changed at all.

	A	C	I_c	L_{geo}	β_L	T_1	T_2
Qubit 1	32 µm	1.2 pF	8.5 µA	168 pH	4.5	13 ns	-
Qubit 2	1 µm	850 fF	1.7 µA	720 pH	3.6	110 ns	105 ns

Table 3.1: Most important design and quantum mechanical parameters.

3.3.2 Experimental setup

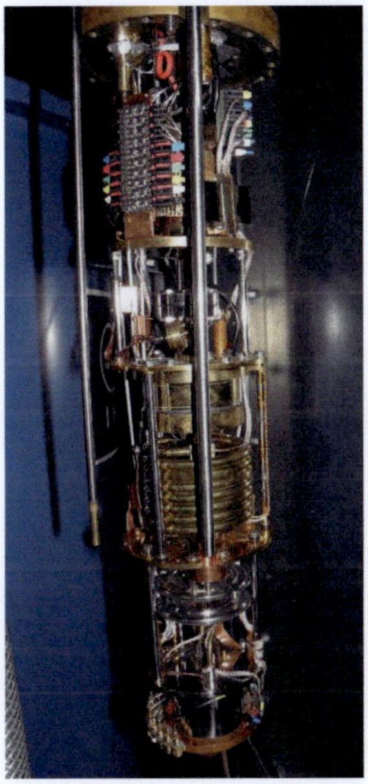

The frequency range for superconducting qubits lies in the microwave spectrum between 5 and 20 GHz. This has two reasons. In principle, going to higher frequencies (about 100 GHz) would be possible. However, the complexity of the setup and the costs for microwave equipment increase drastically for frequencies above $\approx$ 20 GHz. The lower end of the range is determined by temperature. For example, a frequency of 1 GHz corresponds to a temperature of $T = hf/k_B \sim 50\,\text{mK}$. To ensure that the first excited state cannot be populated due to thermal excitation, the operating temperature has to be significantly lower than the excitation energy. Modern Dilution refrigerators (picture on the left, Oxford Kelvinox 100) may reach temperatures down to 10 mK[Ins], in our case about 30 mK. Their mode of operation is comparable to evaporative cooling. At temperatures below 2 K, which can be achieved just by pumping on liquid ^{4}He, a mixture of two isotopes ^{3}He and ^{4}He undergoes a spontaneous phase separation to form a ^{3}He-rich and a ^{3}He-poor phase. Exactly like in an evaporation process, a transition of ^{3}He atoms from the ^{3}He-rich to the ^{3}He-poor phase costs energy resulting in a cooling effect. The volume in which this takes place is known as the mixing chamber. In a continuous circulation process, the ^{3}He/^{4}He-mixture is pumped from the side of the ^{3}He-poor phase, outside the mixing chamber ^{3}He is separated from ^{4}He

(^{3}He has a lower boiling temperature) and delivered back to the ^{3}He-rich phase in the mixing chamber.

The experimental setup is outlined in Fig. 3.4. The green dashed box symbolizes the chip (Sec. 3.3.1), which is cooled to a temperature of ≈ 30 mK. To perform quantum manipulations on the qubit (blue), the continuous microwave generated by the microwave source[2] (µW) passes two mixers[3] connected in series. In the "off"-state, they attenuate the signal sufficiently to ensure free evolution of the quantum system. The DC pulses of the first port of the arbitrary waveform generator[4] (AWG), with the help of the mixers, are used to switch the microwave on and off. Before the microwaves reach the qubit they are attenuated by ≈ 30 dB (as sketched in the purple box in Fig. 3.4) and they pass the on-chip DC-break (dashed orange box) which filters out the low frequency noise.

As discussed in Sec. 3.2.2, the potential energy of the qubit is controlled by applying external flux, which is provided by the current source I_{ext}. The second port of the AWG generates the fine tuning of the flux bias. Furthermore it is responsible for the realization of the flux pulse sequences, as well as the readout pulse. Both signals are combined by a bias tee (bias-T, red) and damped by approximately 30 dB at low temperatures to decrease the noise level and to increase the resolution of the pulse amplitude. To filter out high frequency components, the chip has a series of low-pass LC-filters (light blue). After generating the readout pulse, the state of the qubit is measured with the dc-SQUID (black) 3.2.3. The experimental setup is described in [Lis08] in more detail.

3.3.3 The measurement cycle

The measurement cycle is displayed in Fig. 3.5. It starts and ends with the reset, where the external flux is turned off (or tuned to the reset level) for some time. Thus, the qubit's virtual mass particle falls down into one well and the dc-SQUID, after being measured, has time to get superconducting again. Then, the qubit potential is tuned to the working point by the current source and the pulse sequence of interest, the length of which should not exceed the qubit coherence times by much, may be applied. Each pulse sequence is terminated

[2]Agilent E8257D
[3]Marki M80420LS
[4]Tektronix, AWG7062B

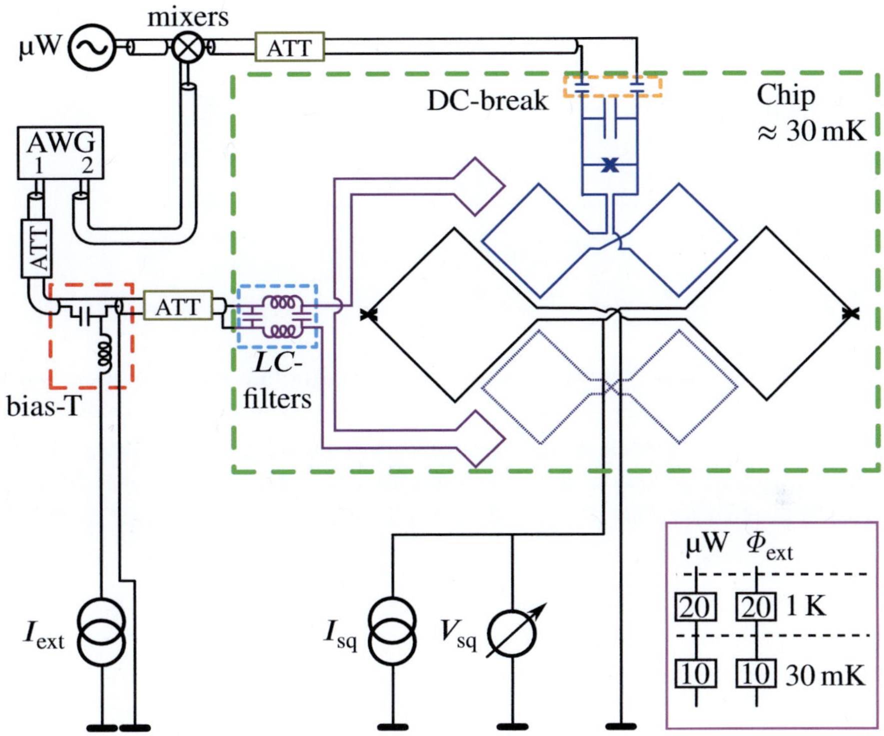

Figure 3.4: Sketch of the experimental setup. The green dashed box symbolizes the chip at a temperature of $\approx 30\,\mathrm{mK}$. The qubit is controlled by microwaves, which are pulsed by using a continuous microwave source (μW) and two mixers in series connected to a channel of our AWG, and by the external flux bias which is partially also controlled by our AWG. The microwave and the flux bias coaxial lines have attenuators at the $1\,\mathrm{K}$ ($-20\,\mathrm{dB}$) and $30\,\mathrm{mK}$ ($-10\,\mathrm{dB}$) levels as skeched in the purple box. The readout is performed by measuring the switching current of the dc-SQUID. At the $1\,\mathrm{K}$ level a current divider (factor 50) and a low pass filter (cut off at $100\,\mathrm{kHz}$) are installed.

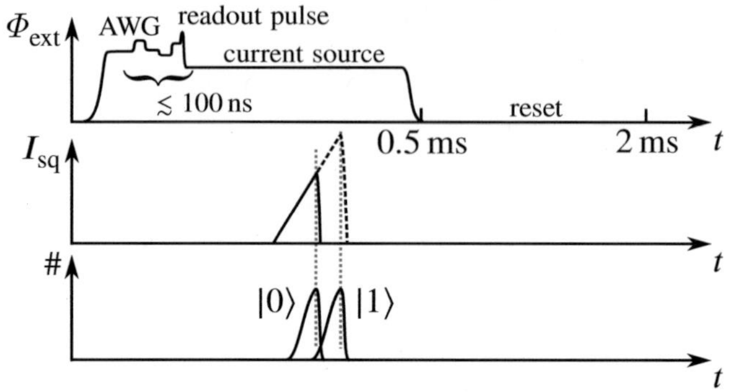

Figure 3.5: The upper plot shows the pulses generated by the current source and AWG. For reading out the qubit, it is parked at a level where both wells are deep enough so that no tunneling occurs. While increasing I_{sq} (middle plot), the time is measured until a voltage drop over the dc-SQUID arises, which results in two histograms, each for the corresponding qubit state $|0\rangle$ and $|1\rangle$ (lower panel).

by the readout pulse. The qubit is then tuned to a stable position where the tunneling between the two qubit wells is turned off, in other words where both qubit wells are deep enough. This allows to determine the switching current of the dc-SQUID and so the qubit state. I_{sq} is increased linearly and the time is measured until a voltage drop occurs. To get a distribution of time intervals, a lot of values (usually between 500 and 2000) are recorded and sorted in a histogram. A time between the two histograms corresponding to the $|0\rangle$ and to the $|1\rangle$ state is chosen as a threshold t_{th} for calculating the qubit population probability as, e.g., $P(|1\rangle) = \#(t < t_{th})/\#_{tot}$ (# denotes the number of events).

3.3.4 Basic experiments for characterizing quantum systems

Excitation spectrum

To find the resonance frequency at a particular flux bias Φ_{ext}, long microwave pulses with varying frequencies f_{μ} are applied to the qubit after which the qubit is read out. The duration of this pulse has to be much longer than the decoherence times of the qubit to ensure that the qubit is in the stationary limit of Rabi oscillations (Sec. 2.1.4). The pulse sequence and the result are shown in Fig. 3.6(a).

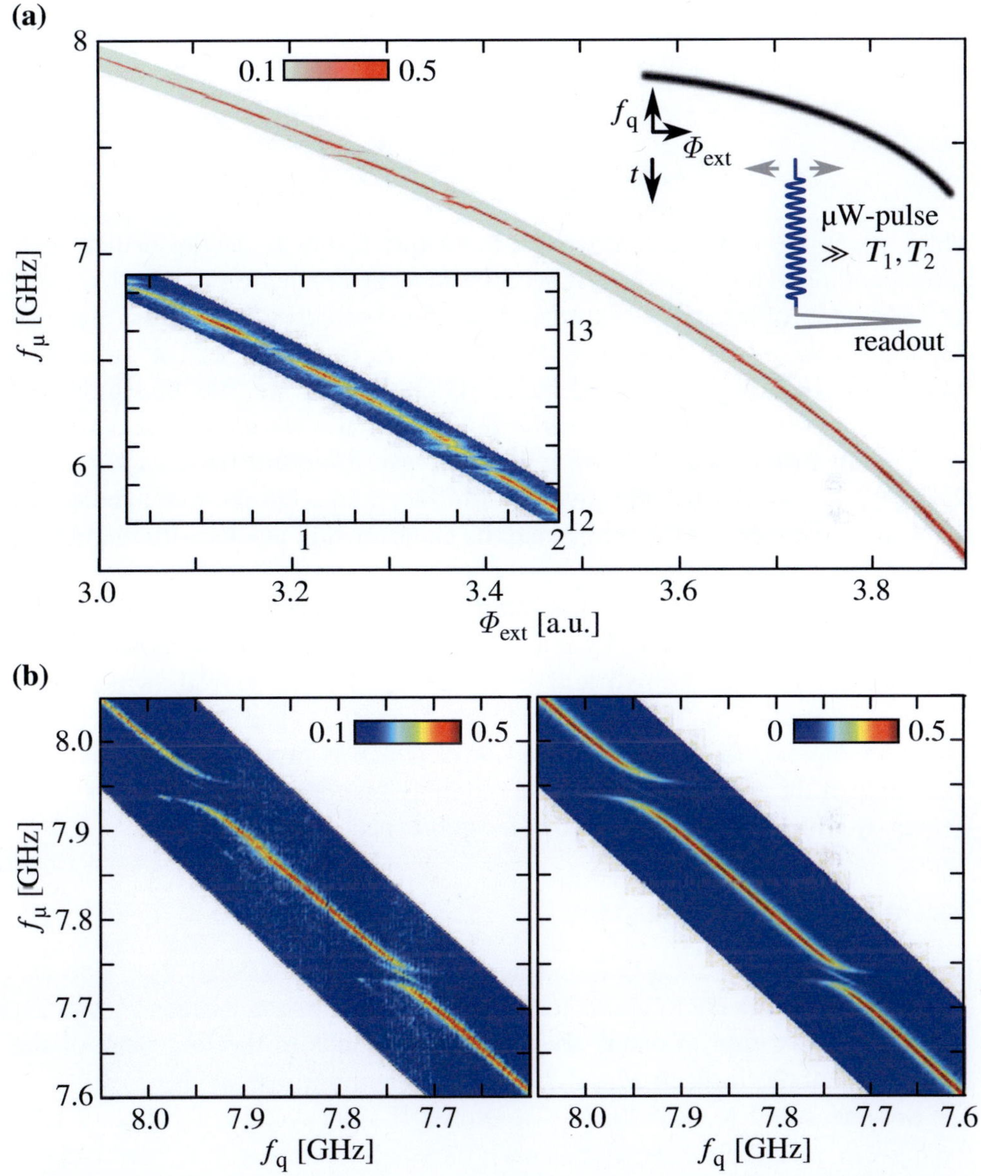

Figure 3.6: Excitation spectrum of the qubit. The upper inset of **(a)** shows the pulse sequence. The upper blurry curve symbolizes the spectrum of the qubit. At different biases (indicated by two grey arrows) long microwave pulses (500 ns) are applied on the qubit and then the readout is performed. A plot of the full excitation spectrum of sample 2 (in the lower inset of sample 1) as a function of the external flux is shown with the excitation probability $P(|1\rangle)$ encoded in color. **(b)** A zoom into the spectrum (with recalculated qubit frequency f_q) reveals two anti-crossings due to coupled TLSs (measurement on the left, simulation on the right).

Since the induced current in the qubit loop is proportional to the externally applied flux, we can express the resonance frequency of the qubit f_q (Eq. (3.11)) as

$$f_q = f_p \sqrt[4]{1 - \left(\frac{\Phi_{ext} - \Phi_{off}}{\Phi_c}\right)^2}, \tag{3.25}$$

where f_p is the plasma frequency of the qubit and Φ_c is the critical flux corresponding to the critical current of the Josephson junction I_c. Φ_{off} accounts for a possible offset of the external flux. In particular, this factor plays an important role when considering the resonance frequency of the qubit as a function of the flux generated by the AWG only. Then, the flux bias produced by the current source tuning the qubit potential to the working point is carried by Φ_{off}. In further analysis, we will rather use the qubit frequency f_q than Φ_{ext}. Φ_{ext} is not only usually given in arbitrary units, but the qubit frequency is also more convenient for comparing the experimental results with theoretical predictions.

At some qubit frequencies, the qubit resonantly interacts with a parasitic system (two-level tunneling system, TLS), and the spectrum shows a level splitting. In Fig. 3.6(b) two anti-crossings can be recognized, which were extensively studied in my diploma thesis [Gra10]. The line width of the spectrum depends on the microwave power which is proportional to the Rabi frequency of the qubit. Thus, in the simulation, one only has to adjust the Rabi frequency which is usually ≈ 5 MHz for qubit spectroscopy.

Decoherence times of the qubit

Once the qubit resonance is found as function of the external flux, one can perform measurements to characterize the qubit's coherence times T_1 and T_2. The π-pulse duration to excite the qubit corresponds to the first peak of the Rabi oscillations (inset in Fig. 3.7(a)). The damping rate of the oscillations $1/\tau$ was measured to be 100 ns. To extract the decay time T_1 (Sec. 2.1.5), the qubit is excited by a π-pulse in each measurement cycle and then read out after waiting for different times t yielding the characteristic exponential decay (Fig. 3.7(a)). The Ramsey fringes (Sec. 2.1.6), which result when applying two $\pi/2$-pulses with varying waiting time t and some detuning of the qubit from the microwave frequency δ in-between, yield the dephasing time T_2 (Fig. 3.7(b)). The pure dephasing time T_2^* can be extracted by using Eq. (2.27). To get reliable values for qubit decoherence times, they have to be measured far away from the

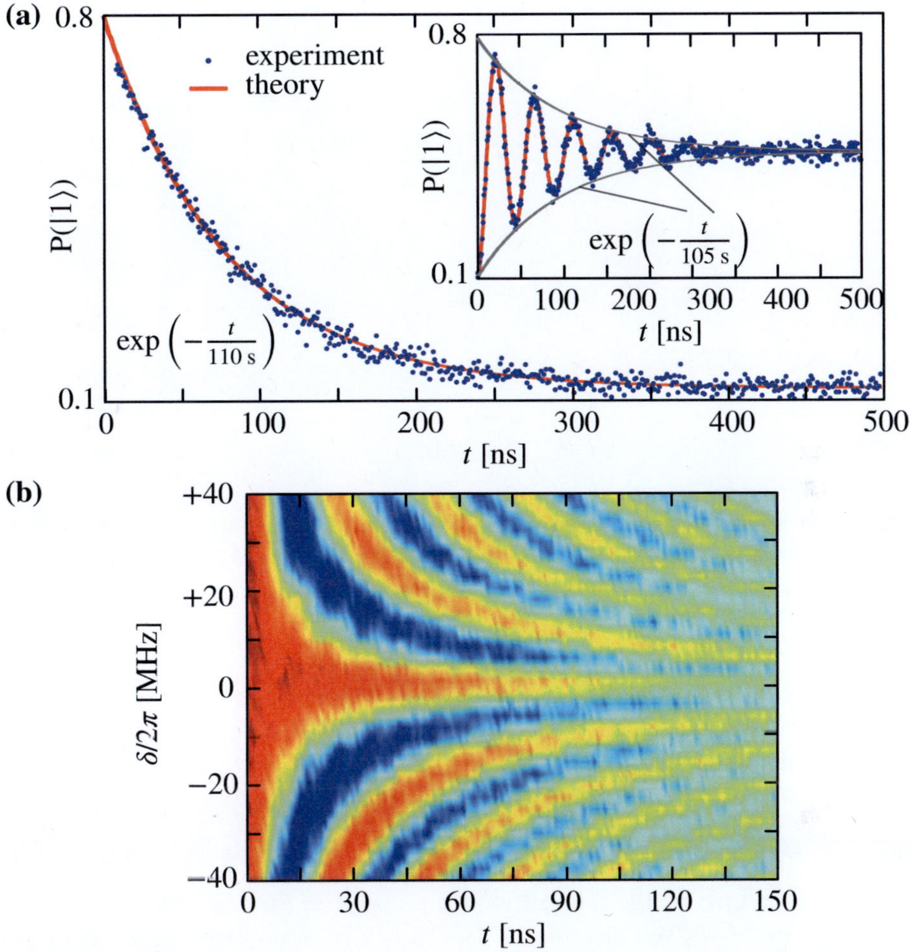

Figure 3.7: Characterization of the qubit. Measurements of the exponential decay and Rabi oscillations **(a)** and Ramsey fringes **(c)** are shown. The simulated curves were renormalized (scaled and shifted) to fit the experiment.

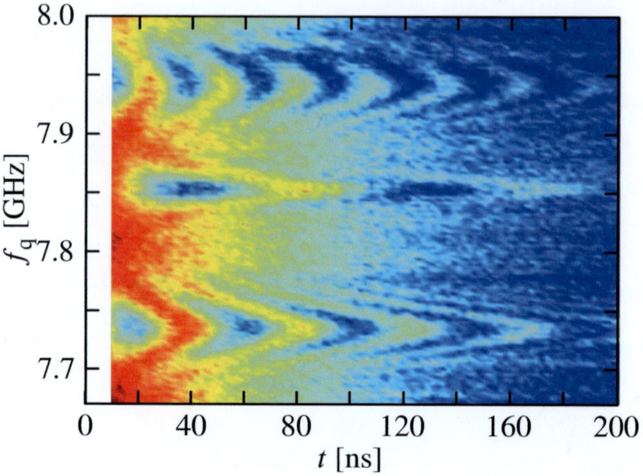

Figure 3.8: Being in resonance with a TLS, the qubit population starts to oscillate between the qubit and the TLS. Close to the TLS resonance, the oscillations are faster but with a smaller amplitude, resulting in a chevron-like pattern.

anti-crossing as they can be very strongly suppressed by the TLS [Mar+05]. The extracted decoherence times T_1 and T_2 of $\approx$ 100 ns are in good agreement with the damping rate of Rabi oscillations (Eq. 2.58).

Decoherence times of TLSs

Since TLSs behave in the same manner as qubits, they are characterized by performing the same experiments. The only difference lies in the excitation and readout. Both procedures can be achieved by swapping the qubit state with that of the TLS, by tuning the qubit in resonance with the TLS of interest and waiting for a full oscillation (Fig. 3.8). It is worth to mention that although there is no other way to readout a TLS, it can be excited directly by the microwaves [Lis+10a]; [Bus+10].

Another very important result of Fig. 3.8 is the obvious sensitivity of the method to detect TLSs. Although the spectrum in Fig. 3.6(b) shows only two TLSs, the time-domain measurement reveals a weakly coupled TLS in-between. This feature will play a key role in the following experiments.

4 Two-level tunneling systems (TLSs)

The tunneling model, which will be discussed in the first section of this chapter, is a phenomenological model developed about 40 years ago. Its aim was to describe the physics behind amorphous solids, or glasses, showing very peculiar and unexpected low temperature characteristics. Although at those times the glass community had only access to macroscopic quantities being average values over a very large number and broad distribution of TLSs, the tunneling model is based on a single tunneling system. Nowadays, it is possible to control quantum systems at the single excitation level. In particular superconducting qubits show coherent interaction with parasitic two-level systems, which is briefly described in the second section. In fact, there are very strong indications that the still existing $1/f$-noise, the limited qubit coherence times, and also parasitic but very coherent quantum systems originate from microscopic tunneling states. Therefore, it is important to verify the tunneling model also on the single TLS level and unmask the physics behind the decoherence effects.

4.1 Tunneling model

4.1.1 A universal model for glasses

Till the year 1971, the thermal characteristics of solids, like the thermal conductivity and heat capacity, were successfully described by the Debye model [EH00] which explained the T^3-dependence of both mentioned quantities at low temperatures below ≈ 1 K. For amorphous solids, no significantly different behavior was expected. This is why the publication of Zeller and Pohl [ZP71] received great attention showing that at low temperatures amorphous quartz has completely different characteristics than its crystalline counterpart (Fig. 4.1). Obviously, glasses have a large number of defects (Fig. 4.2). However, according to the Debye model, single defects should not play a role at low temperatures. Due to the long wavelength of thermal phonons, a point-like atomic defect does not introduce scattering so that the mean free path of phonons is limited by

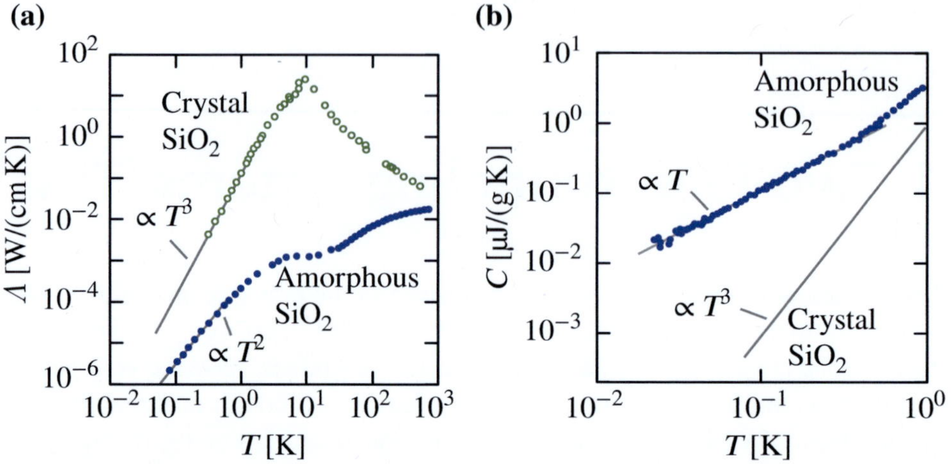

Figure 4.1: According to the Debye model, the thermal conductivity Λ **(a)** and the heat capacity C **(b)** of solids at low temperatures scales as T^3. Point-like defects, due to the long phonon wavelengths, should not play a role. Zeller and Pohl [ZP71] (a) showed the first clear discrepancy between the Debye model and the characteristics of amorphous solids, which, among others, was also confirmed by [Las+75].

the dimensions of the sample. Furthermore, it was completely unclear, how defects should be able to store energy. Another very interesting fact is the universality of such low temperature behavior. It turned out, that also other types of glasses, and even inorganic and organic polymers show almost same characteristics as amorphous silica [EH00]. This lead to the development of a lot of new models [Phi81].

4.1.2 Tunneling systems

Phillips [Phi72], and independently also Anderson *et al.* [AHV72], developed in 1972 the tunneling model. This model has been tested already over four decades and till now it provides correct results for a great variety of experiments on amorphous solids, e.g. phonon scattering, sound velocity, ultrasonic measurements, interactions with external dc and ac electric fields, dielectric characteristics, and last but not least the strain dependence of single quantum systems coupled to the phase qubit studied in this work. Although the tunneling model was a breakthrough, there are also measurements being not fully consistent with the assumptions that are usually made, e.g. [BH77]; [Ska10].

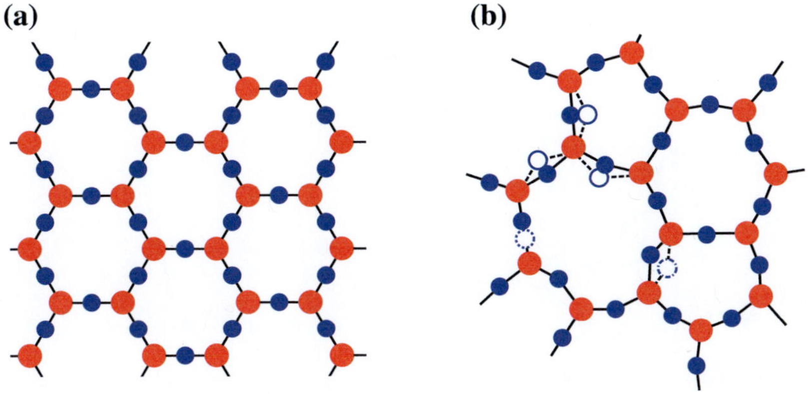

Figure 4.2: (a) A crystalline structure of SiO_2 is shown, with silicon atoms colored in red and oxygen in blue. **(b)** An amorphous glass might have not only "simple" defects, but also atoms, or a small group of atoms, having two stable positions with a nonzero tunneling probability between them. Nevertheless, the stoichiometric coefficients have to stay constant. (Picture taken from [Hun07])

The tunneling model is based on defects, which, however, are not just impurities in the lattice but form bistable systems. Fig. 4.2(b) shows three different possibilities, how amorphous quartz can look like on atomic scale (copied from [Hun07]). Not only tunneling of single atoms between two positions might occur, but even tunneling of a group of atoms is imaginable. It is important to mention that on larger scale, the stoichiometric coefficients stay unchanged, since oxygen atoms still need two neighbors and silicon atoms three. Quantum mechanically, such a two-level tunneling system (TLS) can be described by using a double well potential (Fig. 4.3). The important parameters are on the one hand the asymmetry energy ε, and on the other hand the tunneling mass m, the distance between the two wells d, the barrier height V, and the ground state energy E_0, which are combined to the tunneling energy

$$\Delta = E_0 e^{-\lambda}, \ \lambda = \frac{1}{\hbar} d\sqrt{2mV} \tag{4.1}$$

by using the WKB-approximation. The Hamiltonian for a TLS has then the form

$$H_{\mathrm{TLS}} = \frac{1}{2}\begin{pmatrix} \varepsilon & \Delta \\ \Delta & -\varepsilon \end{pmatrix}. \tag{4.2}$$

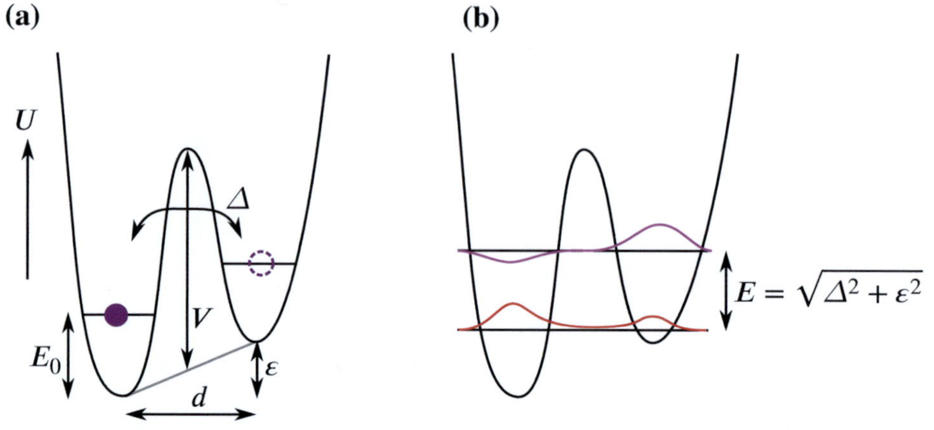

Figure 4.3: (a) According to the tunneling model, a TLS is described by a double well potential with the parameters being the distance between the two wells d, the ground state energy of one well E_0, the barrier height V, the tunneling mass m and the asymmetry energy ε. All parameters except ε can be combined into the tunneling energy Δ (see text), so that the energy splitting of the eigenstates **(b)** of a TLS is $E = \sqrt{\Delta^2 + \varepsilon^2}$.

Diagonalization of this matrix yields for the level splitting of the eigenstates

$$E = \sqrt{\Delta^2 + \varepsilon^2}. \tag{4.3}$$

Obviously, all tunneling systems in an amorphous material are different so that for a macroscopic analysis a concrete distribution of the parameters is required. The success of the tunneling model is based on two assumptions [Phi81]. The first one is to take the independent parameters ε and λ for the distribution and to assume that the distribution is constant in the intervals $d\varepsilon$ and $d\lambda$, i.e.

$$P(\varepsilon, \lambda) = P_0 \, d\varepsilon \, d\lambda. \tag{4.4}$$

The second assumptions concerns the way of coupling between the TLSs and their environment, which is briefly discussed in the following.

4.1.3 Coupling of TLSs to their environment

Being local defects, the TLSs couple to local strain, e.g. phonons or static deformation of the sample. In such experiments, e.g., the phonon scattering or the ultrasonic absorption were studied [Cla+94]. Furthermore, charged TLSs interact also with external electric fields due to their dipole character. The interesting values in that case are, e.g., resonance measurements or dielectric losses [VH77]. In particular in the case of elastic fields, a change of basically all parameters of the double well potential (Fig. 4.3) is expected. However, measurements suggest that the change of the asymmetry energy dominates the modification of Δ by much [EH00]. In principle, the reason is the large factor λ, whose variations are exponentially suppressed. Therefore, the corresponding perturbation Hamiltonian reads

$$H_{\mathrm{per}} = \frac{1}{2}\begin{pmatrix} \delta\varepsilon & 0 \\ 0 & -\delta\varepsilon \end{pmatrix}. \tag{4.5}$$

The variation of ε scales as

$$\delta\varepsilon = 2\gamma\varepsilon_{\mathrm{d}} \tag{4.6}$$

where ε_{d} denotes the mechanical deformation, e.g. due to phonons or external strain, and γ is the so-called deformation potential of the respective TLS describing its strength of the strain dependence with a rough magnitude of 1 eV [BH77]. Taking into account the perturbation, the function of the TLS excitation energy (Eq. 4.3) has the form

$$E = \sqrt{\Delta^2 + (\varepsilon + \delta\varepsilon)^2}. \tag{4.7}$$

On the other hand, assuming the TLS to be a charged dipole, the asymmetry energy can be also controlled by external electric fields. Then, $\delta\varepsilon = 2\vec{p}\vec{E}$ with $\vec{p}$ and $\vec{E}$ being the TLS electric dipole moment and the externally applied electric field, respectively. In our case, the TLS dipole moment plays in so far an important role as it is responsible for coupling the TLS to the qubit.

4.1.4 An example tunneling system: KCl:Li

To get a more concrete image of tunneling systems, in the following an artificial and very well studied example is presented. Harrison *et al.*, under supervision of Pohl, investigated the characteristics of the salt KCl doped with Li^+-ions (KCL:Li). They published their results in [HPP68], which is also summarized in [EH00].

A cross-section of the crystal KCl is schematically shown in Fig. 4.4(a). The Li^+-ion is smaller than K^+-ion and has therefore eight equal corners of a cube

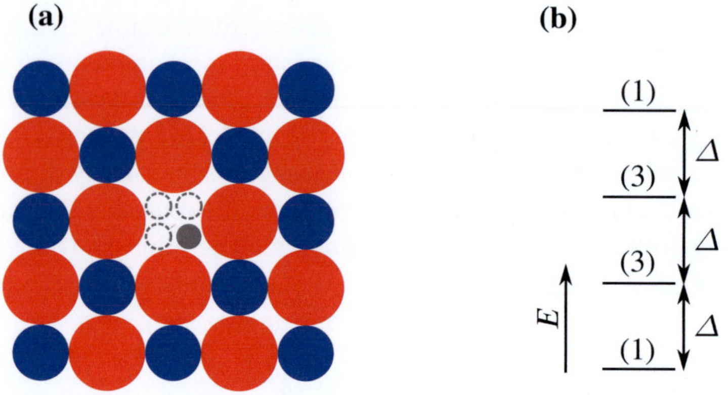

Figure 4.4: (a) The KCl lattice is shown (K^+ in blue and Cl^- in red) with a K^+-ion replaced by a small Li^+-ion. The ion sizes on the picture correspond to real proportions: $r_{Cl^-} = 1.81\,\text{Å}$, $r_{K^+} = 1.33\,\text{Å}$ and $r_{Li^+} = 0.6\,\text{Å}$. **(b)** This results in eight eigenstates, where the first and the second excited states are triply degenerated (see text). Among this, the levels are separated equidistantly.

to occupy. From the measurements shown in Fig. 4.5 the main parameters of this eight-well potential can be extracted. The tunneling distance of the Li-ions along an edge is 1.4 Å, which is a bit less than half the lattice constant, while the barrier height corresponds to $V/k_B = 100\,\text{K}$. Due to symmetry reasons, the eight states have only four different energies. The ground and the highest excited states, which correspond to a superposition of all wells with the same phases and with π-shifts between the neighboring wells, respectively, are non-degenerated. The other two states are triply degenerated, where either the phases of opposite sides (first excited state) or of neighboring edges (second excited state) of the cube differ by π. Apart from degeneracy, the energies of the eigenstates differ by the same amount, so that the results of this model resemble strongly those of a double-well potential. The energy splitting is estimated to be $\Delta = 10^{-4}$ eV which corresponds to 24 GHz. Such a tunneling system has obviously a large electric dipole moment, which was also extensively investigated. For example, in external electric fields the degeneracy is lifted.

Furthermore, this experiment allows one to also investigate the tunneling behavior as a function of the tunneling mass. Harrison *et al.* used two isotopes of Li: ^{6}Li and ^{7}Li. The result is shown in Fig. 4.5(b). The heavier ions have a larger contribution to the specific heat. By rescaling the temperature axes, the points of both measurements can be ``tuned'' to lie on one curve. One can see that the theoretical curve then fits both isotopes perfectly.

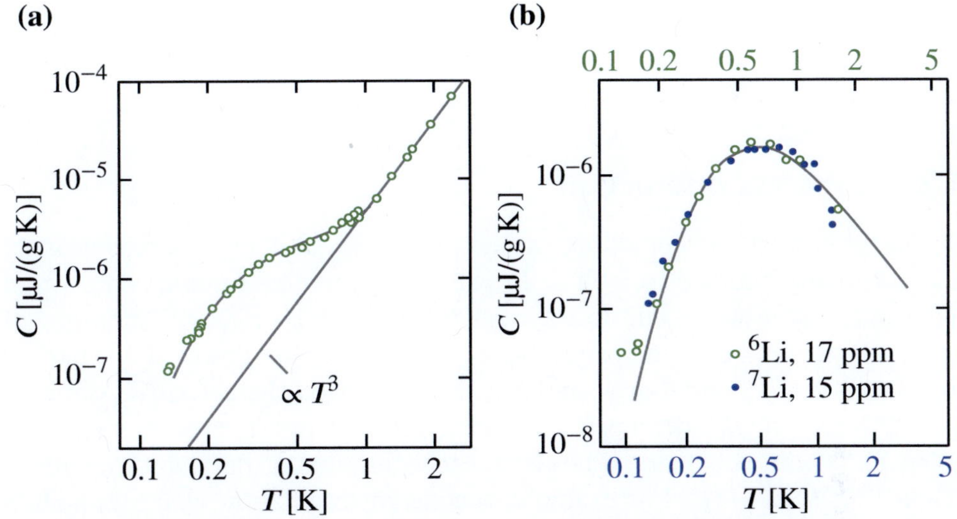

Figure 4.5: (a) Below 1 K the tunneling Li^+-ions start to dominate the phonon contribution to the specific heat that has a T^3 dependence on temperature in undoped KCl salt. **(b)** The points and the theory corresponds to the difference between the specific heat of doped and undoped KCl, in other words to the difference between the two curves in (a). Furthermore, also the different contributions to the specific heat of the two isotopes $^6Li^+$ and the slightly heavier $^7Li^+$ were investigated.

4.2 A TLS seen by the qubit

4.2.1 Incoherent interaction

Superconducting qubits are promising candidates for being used in a sold-state quantum computer. While a decade ago only qubit resonances could be observed [NPT99]; [Mar+02], nowadays the coherence times exceed the duration of an elementary quantum gate by many hundred times. This quite rapid progress was not only achieved by clever engineering, but also by an intensive material research [Kli+09]; [Ska10]. In future, the defect states in the used materials will be the limiting factor for further improvement of the qubit coherence [Pal+13]. The confidence is growing that incoherent TLSs, or fluctuators, cause the mysterious $1/f$-noise [Shn+05]. This type of noise is present in a very large variety of systems suggesting to be ubiquitous [Pre78]. For the superconducting qubits, the most important sources for decoherence effects are charge and flux noise and noise of the critical current of the Josephson junction, as well as dielectric losses [Pal+13], all of which can be explained by spurious defects located inside the junction or on the surfaces of the substrate and the superconductor. The decoherence of the qubit-TLS system was extensively studied in [MSM09].

4.2.2 Coherent interaction

Beside the incoherent interaction, a superconducting qubit might be also strongly and coherently coupled to a TLS. In particular, the phase qubits suffer from these parasitic systems due to their relatively large JJ. First observations of anti-crossings in the qubit spectra were published by the group of J. Martinis in 2004 [Sim+04], and four years later, with improved characteristics of phase qubits, Neely *et al.* demonstrated full control of the TLS state [Nee+08] using TLSs with coherence times even exceeding that of the qubit [MSM09]; [Lis+10b]; [Gra+11]. For example, a coherent oscillation of the excitation between the qubit and a TLS is plotted in Fig. 4.6(b). In agreement with the tunneling model is the fact that the TLS distribution is reset when the sample is warmed up to about 20 K, on the one hand allowing to study a TLS ensemble with only one sample [Sha+10], on the other hand, however, making it significantly difficult or impossible to use them for quantum computing. The most concrete model of a TLS was provided by Cole *et al.* assuming oxygen atoms tunneling inside the amorphous $AlO_{1.25}$ layer [DuB+13]. Fig. 4.6(a) shows the ground state in such a case.

4.2.3 Alternative models

Beside the tunneling model, there are also other models phenomenologically explaining the anti-crossings in the qubit spectra. Each theory describing a potential quantum system interacting with the phase qubit has to suggest a coupling mechanism. Therefore, we have to look at the full Hamiltonian of superconducting qubits:

$$H_{\mathrm{q}} = E_{\mathrm{C}}(\hat{q} - n_{\mathrm{G}})^2 + E_{\mathrm{L}}(\hat{\varphi} - \varphi_{\mathrm{ext}})^2 - E_{\mathrm{J}} \cos \varphi. \tag{4.8}$$

The Hamiltonian consists of a harmonic potential modulated by the cosine term. The conjugated variables are still the charge operator $\hat{q}$ and the Josephson phase $\hat{\varphi}$ fulfilling $[\hat{q}, \hat{\varphi}] = i\hbar$. The first term corresponds to the charging energy which can be controlled by a gate electrode n_{G}, the second term is responsible for the magnetic energy which depends on the external normalized flux φ_{ext} and the

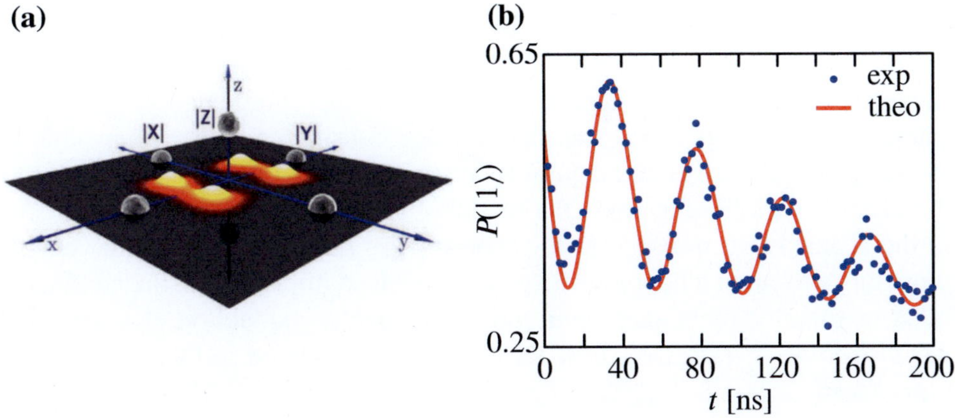

Figure 4.6: (a) A "photograph" of a possible TLS ground state which is formed by a tunneling oxygen atom inside the JJ with a stoichiometry of $AlO_{1.25}$ [DuB+13]. **(b)** Observed and fitted coherent oscillations between the phase qubit and a TLS. The amplitude of the oscillations (red curve) can be derived analytically [MSM09].

last part determines the energy of the Josephson junction (JJ). The idea is now to think about physical systems which might couple to one or to several of these three terms. All theories have in common that they are based on microscopic defects. A detailed analysis of experimental results could not completely exclude some theories, however, it provided a restriction on the coupling mechanism by confirming a basically zero longitudinal coupling [Col+10]). Therefore, the effective interaction Hamiltonian between the qubit and the TLS can be assumed to have only a perpendicular component (see also Sec. 5.3.1) being responsible for the coherent energy oscillations (Figs. 3.8 and 4.6(b)). In the case of coupling to electric fields, this condition is fulfilled right from the start.

The most obvious coupling mechanism is the electric interaction, predominantly inside the JJ due to the strong electric fields required for a coupling. While in this thesis we consider tunneling of charged atoms, Agarwal *et al.* recently published a polaronic model of two-level systems suggesting that all TLSs in amorphous solids observed so far are actually electrons dressed by optical frequency phonons tunneling between atoms [Aga+13]. Other possible candidates for the parasitic systems are entangling bonds or Andreev bound states [Sou+09]. For the latter two models it has to be investigated whether and how they couple to strain fields. The flux noise might be introduced by a large number of free spins on the surface of the superconductor [Sen+08]. However, due to the large magnetic moment required for an appropriate interaction strength

(10^5 electron spins, [Col+10]), this idea can be ruled out for explaining coherent interaction with the phase qubit. The last possibility to couple to the qubit is via modulation of the critical current of the JJ, in other words coupling to $\cos\varphi$. Interestingly, this approach was the first one considered to be responsible for inducing anti-crossings in the qubit spectrum [Sim+04]; [FI07]. To understand the principle, we have to apply the model of superconducting channels existing in the JJ making it transparent for cooper pairs [KY05]. Then, either again a tunneling system or a hopping electron between an impurity and the conduction band of the electrodes can open and close such a conduction channel. This has also to be investigated, how far the results presented in this thesis can be reproduced.

5 Experimental

The aim of my PhD thesis was to investigate the behavior of two-level systems (TLSs, Sec. 4) coupled to a superconducting phase qubit (Sec. 3) as a function of the externally applied mechanical strain and thus verify the tunneling model [Phi72]; [AHV72]. In the first section, all necessary parts and pre-calculations are presented concerning the bending of the chip. The two spectroscopy techniques used for the measurements are explained in detail in the second section. By chance, in one spectrum we found a very peculiar, non-linear and non-hyperbolic TLS trace, which could be readily qualitatively and quantitatively explained and simulated by assuming two TLSs interacting coherently with each other via dipole-dipole coupling. The full description of the formed tripartite system is presented in the third section.

5.1 Preparations

5.1.1 Sample holder

In order to bend the chip with a piezo actuator, we firstly had to construct a completely new sample holder. It is shown in Fig. 5.1 and described in much more detail (including engineering drawings) in [Pei12]. The main part of the sample holder consists of two pieces: the carrier plate with a notch for the chip and the carrier housing where the piezo is mounted. The chip is embedded into the notch ''face down'' and fixed from the side of the housing with a self-adhesive copper foil. The piezo is glued onto a support plate which can be displaced very accurately by a fine-thread screw. To minimize the shear forces on the piezo and to have a well defined point on the chip on which the force generated by the piezo is acting on a gold plated zirconia ball was glued on the piezo.

According to the datasheet, the used piezo actuator[1] has a nominal displacement of 65 nm/V in the approved voltage range between −30 and 120 V. Being

[1]PICMA®Piezoelectric Stack Actuators, P-882.11

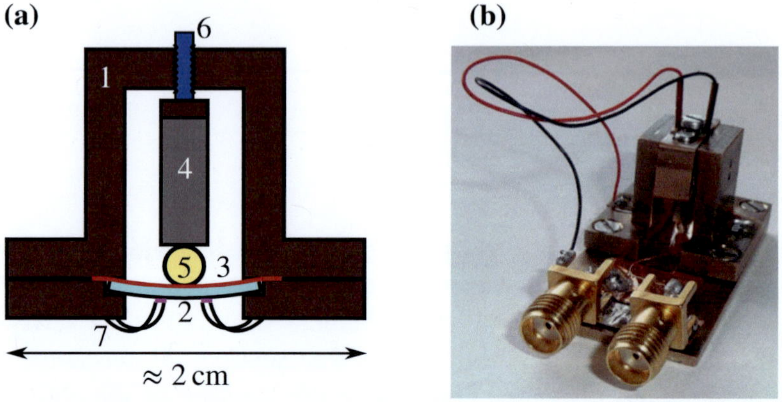

Figure 5.1: (a) Schematic of the sample holder. (1) The carrier material is a copper-beryllium alloy. The chip with the qubit (2) is held by a notch from one side and by a self-adhesive copper foil (3) from the other side. The chip is bent with the help of the piezo actuator (4) whose force is mediated to practically a single point on the backside of the chip via the zirconia ball (5). For accurate positioning (see text), the piezo can be shifted by a screw (6). Both qubits on the chip (purple rectangles) are connected to the lines via bonding wires (7) through a hole in the carrier plate. An image of the sample holder is shown in **(b)**. The SMA-connectors were mounted only for the determination of the piezo constant via capacitance measurements.

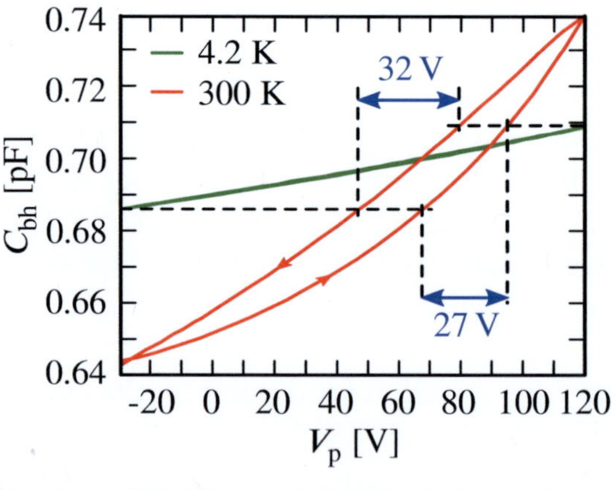

Figure 5.2: The calibration of the piezo actuator at $4.2\,\mathrm{K}$. Measuring and comparing the voltage ranges for corresponding capacitances formed by the gold plated ball and the housing yields a reduction factor of ≈ 5 of the piezo constant between room and liquid helium temperature.

not in contact, the gold plated ball and the copper foil on the carrier plate together with the housing form the capacitor C_{bh} which allows the characterization of the piezo at low temperatures. Fig. 5.1(b) shows the sample holder modified specially for this type of measurement with the SMA connectors used to connect the ball-housing capacitor to the high frequency signal of the capacitance meter[2]. At room temperature, the initial distance between the ball and the plate can be easily adjusted and the capacitance curve shows also the expected hysteresis of the actuator (Fig. 5.2). While cooling down to low temperatures, however, it turns out that the expansion coefficient of the metal is much higher than that of the piezo stack (will be discussed some lines below) yielding very rapidly a contact between the housing and the ball. Therefore, several cool downs are required until the distance between the ball and the carrier plate is adjusted via the fine-thread screw so that the capacitance at low temperature is in a proper range. The capacitance values for the whole piezo range (Fig. 5.2) of 150 V at 4.2 K correspond to a range of only ≈ 30 V at room temperature yielding a factor of 5, which is also in correspondence with [Tay+06]. Taylor *et al.* measured a piezo actuator of the same type but with slightly different dimensions and, although they decreased the temperature only down to 40 K, a linear extrapolation of their results to 4.2 K yields exactly the factor 5. Since we do not expect the piezo characteristics to change significantly at 30 mK, the value of the piezo constant is evaluated to be 13 nm/V. Furthermore, Taylor *et al.* [Tay+06] also measured the expansion coefficients of copper and their piezo actuator directly. Again, extrapolating their curves[3] down to very low temperatures yields a shrinking of 33 μm for a 9 mm bulk of copper, while the length of the piezo ceramic changes only by 11 μm. The difference is about 22 μm. This value is in agreement with our observations. Even when adjusting the distance to the maximal verifiable value of 10 μm between the ball and the copper foil[4], the ball comes into contact with the copper foil very quickly after starting to cool down the setup.

Before cooling down the sample holder with the qubit chip to the base temperature of 30 mK, we adjust the ball to be in contact with the copper foil. This is the easiest way to ensure a permanent contact between the chip and the

---

[2]Andeen Hagerling, 2500A 1 kHz ultra-precision capacitance bridge

[3]Linear approximation of the expansion coefficients of the data in [Tay+06]: copper: $\alpha_{Cu}(T) = \{0,\ T < 14\,\text{K};\ 0.13 \cdot 10^{-6}/\text{K} \cdot T - 1.8 \cdot 10^{-6},\ 14\,\text{K} < T < 140\,\text{K};\ 16.5 \cdot 10^{-6}/\text{K},\ 140\,\text{K} < T < 300\,\text{K}\}$; piezo stack: $\alpha_p(T) = 0.021 \cdot 10^{-6}/\text{K} \cdot T + 0.66 \cdot 10^{-6}$.

[4]In the maximal voltage range between −30 and 120 V the maximal displacement is about 10 μm at room temperature.

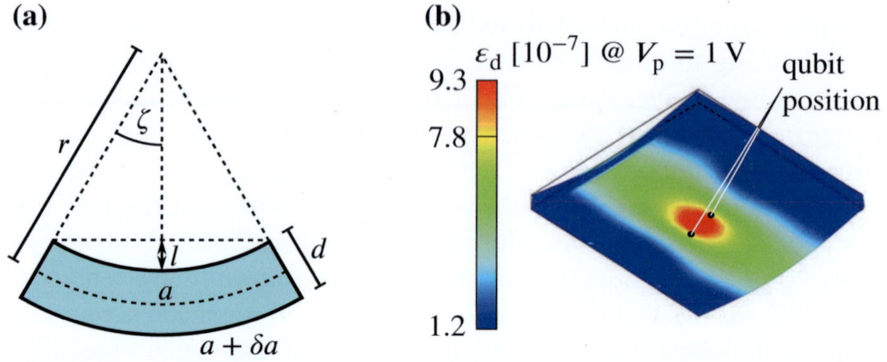

Figure 5.3: **(a)** Geometrical considerations allow to get a feeling for the chip deformation. Approximating the chip surface to have a uniform deformation (like bent on a cylinder), the very small ζ and $r \gg a$ yield a strain of $\varepsilon_\mathrm{d} = 3 \cdot 10^{-7}$ at $1\,\mathrm{V}$ on the piezo ($l = 13\,\mathrm{nm}$). This corresponds to a bending radius of $r = 370\,\mathrm{m}$ and to an angle of $\zeta = 5 \cdot 10^{-4\circ}$ (see text). **(b)** Numerical simulation of the bending process yields a more precise value of $\varepsilon_\mathrm{d} = 7.8 \cdot 10^{-7}$, which is in the same order of magnitude.

ball. Therefore, we can expect the chip to be initially pre-bent by $\approx 20\,\mu\mathrm{m}$, however, for the experiment itself, this is completely irrelevant since it is just a constant offset in strain.

5.1.2 Bending of the chip

Fig. 5.3(a) shows an exaggerated sketch of the bent chip. One can easily make rough estimations of the deformation by assuming the chip surface to have a cylindrical shape, because the notch is holding the chip at two opposite sites. The size of the chip, which corresponds to the middle line, is $a = 6.2\,\mathrm{mm}$ and it has a thickness of $d = 400\,\mu\mathrm{m}$. When displacing the center of the chip by l from its resting position, the chip surface gets elongated by δa. The bending angle ζ and radius r as a function of a and l are evaluated by using the two equations

$$a = 2\zeta r \quad \text{and} \quad l = r - r\cos\zeta \approx \frac{1}{2} r \zeta^2 \tag{5.1}$$

yielding

$$r = \frac{a^2}{8l} \quad \text{and} \quad \zeta = \frac{4l}{a}. \tag{5.2}$$

The deformation coefficient ε_d for such a simple case can be expressed as

$$\varepsilon_\mathrm{d} = \frac{\delta a}{a} = \frac{2\zeta(r + 0.5\,d) - 2\zeta r}{a} = \frac{\zeta d}{a} = \frac{4ld}{a^2} \overset{l=13\,\mathrm{nm}}{\approx} 3 \cdot 10^{-7}\frac{V_\mathrm{p}}{\mathrm{V}}. \tag{5.3}$$

For each Volt on the piezo, l increases by 13 nm resulting in an enhancement of the deformation ε_d by $3 \cdot 10^{-7}$.

A much better estimation of the deformation parameters is, however, a finite element simulation. We used the engineering program Autodesk Inventor to model the chip bending as a stress analysis. The model of our sample holder was implemented into the program as close to reality as possible, including the design, the zirconia ball and, of course, the corresponding parameters of all used materials. The result is shown in Fig. 5.3(b). The displacement of the chip center from the point of origin corresponds again to 1 V on the piezo actuator, which is 13 nm. The deformation profile has an elliptic-like form. The two black dots on the chip symbolize the position of the qubits (or JJs). At these places the local strain results to be $7.8 \cdot 10^{-7}$ being in the same order as the geometrically estimated value. Another very interesting outcome of the simulation is the spring constant of the chip, in other words the back acting force on the piezo. In the case of this force being not negligible, the piezo displacement would be strongly affected. At room temperature, the piezo stiffness [PI], defined as the generated maximal force over maximal displacement, is 24 N/μm for our piezo. Compared to this, the simulated spring constant of the silicon chip of 0.7 N/μm is indeed negligible. Unfortunately, it is difficult to find the piezo characteristics at low temperatures. Assuming that the acting force of the piezo decreases in the same manner as the displacement [Pei12], the stiffness remains more or less the same and the back action of the chip can be neglected.

5.2 Tuning the TLSs

After constructing and analyzing the experimental features of the sample holder, we could test whether the TLSs feel the generated strain and change their resonance frequencies. The first measurements were performed with sample 1 (Sec. 3.3.1) due to its high number of TLSs. We could already observe first hints for the shift of the TLS resonance frequencies by recording standard spectra as in Fig. 3.6. By using the so called multi photon spectroscopy protocol (MPS), we succeeded in taking the TLS traces more systematically. The rather low coherence times of sample 1 ($\approx$ 13 ns) did not allow us to also perform

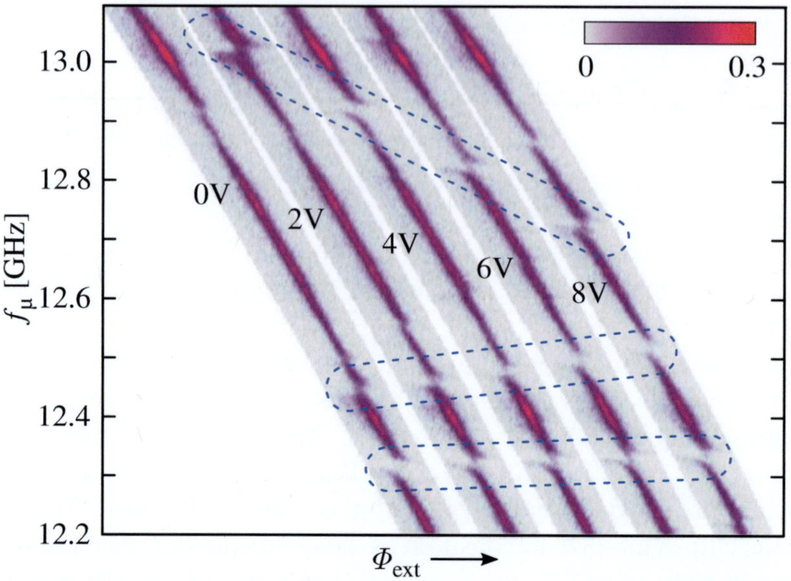

Figure 5.4: Full qubit spectra recorded at different piezo voltages and shifted horizontally for clarity. Three anti-crossings moving with strain are marked by blue rectangles.

time domain experiments. However, by using a more coherent qubit ($\approx$ 100 ns, Sec. 3.3.1), we developed a new technique which is much more sensitive and significantly decreases the measuring time (single photon spectroscopy, SPS).

5.2.1 Multi Photon Spectroscopy (MPS)

The most obvious way to verify whether the TLSs feel the changing strain is to observe the moving anti-crossings in the qubit spectrum. The spectra were recorded as explained in Sec. 3.3.4 at several piezo voltages. In Fig. 5.4 some of such measurements are shown which are shifted horizontally for better visibility. Three moving anti-crossings are marked showing a random direction and strain dependence already at this stage.

In principle, it is superfluous to record a whole spectrum at each voltage. Instead, only the information whether there is a TLS at a frequency or not is interesting. Therefore, we fit the qubit resonance curve as a function of the external flux by the theoretically derived formula in Eq. (3.25), and try to excite the qubit at each flux, and voltage, only at the expected resonance frequency. Fig. 5.5 visualizes this idea. The points along the yellow curve are

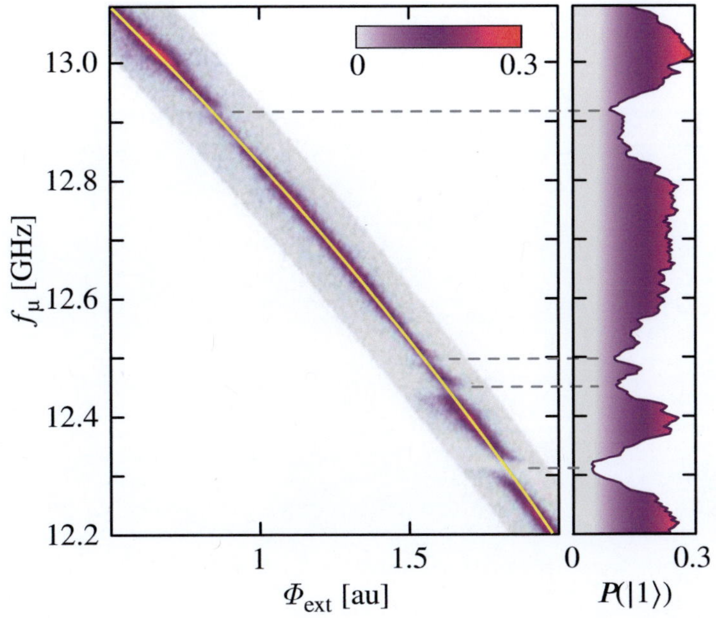

Figure 5.5: For the MPS, only the points along the expected qubit resonance curve (yellow) are measured, or, as shown on the plot, extracted from the full spectrum. In such a curve, the anti-crossings are represented as dips, because in the middle of the anti-crossing the qubit-TLS system does not get excited. Measuring and merging such curves into a 3D plot yield the MPS.

extracted from the spectrum and plotted on the right side of the figure. The TLS resonances, being anti-crossings on the spectrum, now turn into dips in $P(|1\rangle)$. The reason is that in absence of a TLS, the qubit gets excited normally by a resonant microwave pulse and yields a corresponding excitation probability. In resonance with a TLS, however, due to the level repulsion the qubit-TLS system is not excited and the qubit population is therefore nearly zero. Merging such curves into a 3D plot with the piezo voltage as the horizontal axis yields an overview image as shown in Fig. 5.6, where the TLS resonances appear in bright color. Such a measurement is referred to as the multi photon spectrum (MPS).

Several key features of the tunneling model (Sec. 4.1) can be clearly recognized in Fig. 5.6 [Gra+12]; [Pei12]. First of all, there are a lot of linear traces, some of which reveal very weakly coupled TLSs which cannot be identified in a standard spectrum as such. Linear frequency dependences arise in the case when the tunneling energy Δ is much smaller than the asymmetry energy ε (the other way around, $\varepsilon \ll \Delta$, results in completely horizontal traces, which are, however,

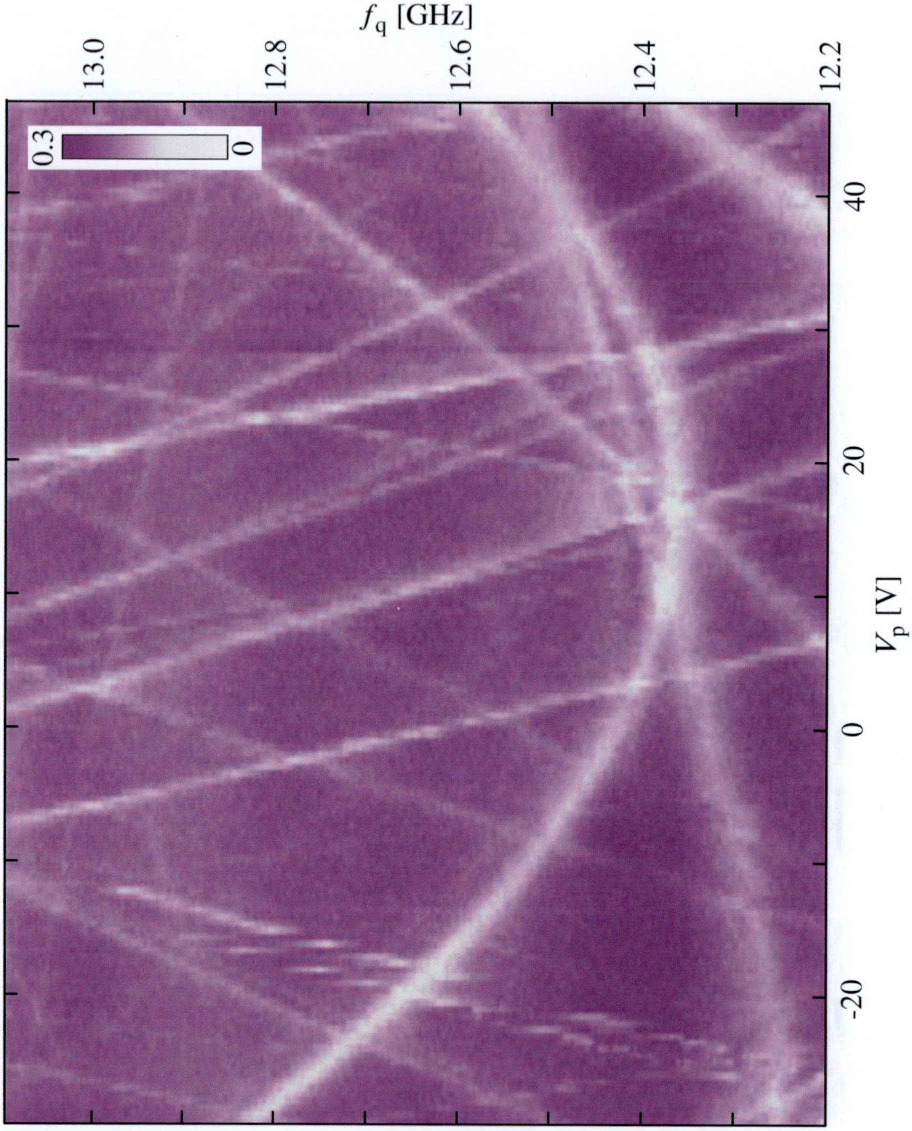

Figure 5.6: In the MPS, the resonance frequencies of TLSs are revealed by bright traces indicating lower qubit population ($P(|1\rangle)$ encoded in color). Most of the traces have a linear strain dependence, which is the case if $\Delta \ll \varepsilon$. On the left of the plot a TLS can be recognized the resonance frequency of which is randomly jumping between two values (see text). The most important feature, however, is the hyperbolic trace where $\varepsilon \lesssim \Delta$.

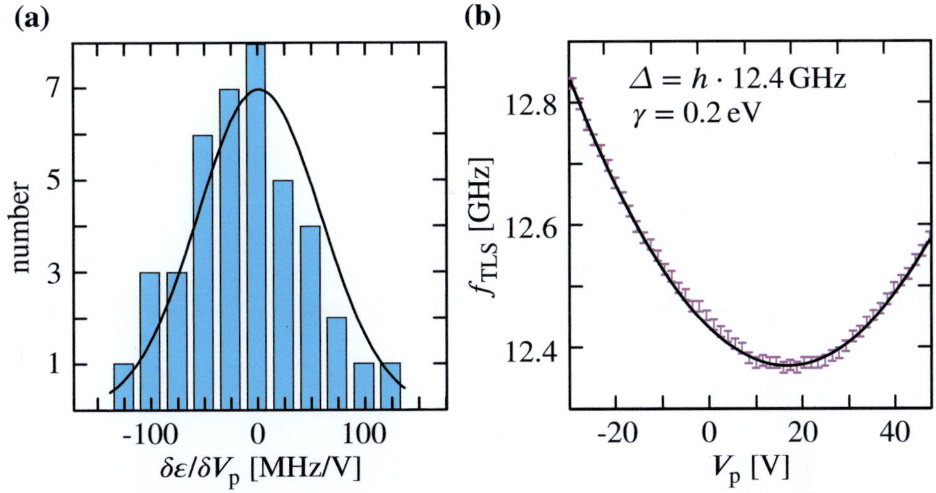

Figure 5.7: (a) A histogram of the asymmetry energy dependence on piezo voltage of 41 TLSs. The shape of the histogram strongly resembles a Gaussian distribution (solid curve). **(b)** The extracted frequencies and the fit to the hyperbolic TLS on Fig. 5.6 is plotted (see text).

hardly observed). In Fig. 5.7(a) a histogram of 41 TLSs is shown covering a range of the deformation potential(Eq. 4.6) of

$$|\gamma| = \frac{1}{2}\left|\frac{\delta\varepsilon}{\delta\varepsilon_\mathrm{d}}\right| \lesssim 0.3\,\mathrm{eV}. \tag{5.4}$$

This number is slightly smaller than expected in glasses ($\gamma \approx 1\,\mathrm{eV}$) [BH77]. However, it is of the same order of magnitude. One reason for the different values might be that TLSs with larger deformation potentials are strongly asymmetric and therefore coupled to the qubit more weakly (Sec. 5.3.3). Furthermore, the number of visible TLSs in our junction is about 10 per GHz, so that the TLS density is about two orders of magnitude smaller than the expected number in glasses [Pei12]. Another very interesting feature is a TLS possessing telegraphic noise of its resonance frequency dependence around the piezo voltage of −20 V. Such structures are predicted by the tunneling model due to a possible direct interaction between TLSs [AH75]. A coherent TLS coupled to the qubit might interact with another TLS jumping incoherently from one well to the other and thus changing the excitation energy of the visible one. In this case, the average resting time is about one hour. An example of a coherent TLS-TLS coupling is presented in the next section. The last but not least striking feature is the broad, nonlinear TLS trace extending over the whole plot. One can extract its resonance frequencies (Fig. 5.7(b)) and fit them with the hyperbolic function (Eq. (4.7))

$$f_{\mathrm{TLS}} = \frac{1}{h}\sqrt{\Delta^2 + (2\gamma\varepsilon_{\mathrm{d}}(V_{\mathrm{p}}))}, \tag{5.5}$$

where

$$\varepsilon_{\mathrm{d}}(V_{\mathrm{p}}) = \frac{7.8 \cdot 10^{-7}}{\mathrm{V}}(V_{\mathrm{p}} - V_0) \tag{5.6}$$

with V_0 being the voltage offset. The fit yields $\Delta/h = 12.4\,\mathrm{GHz}$ for the tunneling frequency and $\gamma = 0.2\,\mathrm{eV}$ for the deformation potential.

5.2.2 Single Photon Spectroscopy (SPS)

A more coherent phase qubit (sample 2, Sec. 3.3.1) opened a door to time domain measurements. It allowed us not only to verify quantum protocols like Ramsey fringes (Fig. 3.7) or echo sequences [Hah50], but also to observe coherent oscillations with strongly coupled TLSs (Fig. 3.8) and even implement more complex protocols, for example entangling two TLSs [Gra10]. In particular, to detect TLSs, measuring the time domain oscillations is a much more sensitive way than recording a spectrum [Sha+10]. The reason is that the line width of the qubit resonance of $\sim 1/T_1 \approx 10\,\mathrm{MHz}$ limits the visibility of anti-crossings. For example, the weakly coupled TLS being responsible for oscillations in the middle of Fig. 3.8 of $\approx 12\,\mathrm{MHz}$, can hardly be recognized on the corresponding spectrum (Fig. 3.6(b)). One further disadvantage of the MPS is the high sensitivity to qubit resonance drifts, so that a time consuming calibration procedure has to be repeated several times to get accurate results.

A significant improvement can be achieved by using the technique of the single photon spectroscopy (SPS). The pulse sequence is shown in Fig. 5.8(a). After exciting the qubit with a π-pulse at an energy preferably outside the region of interest, the qubit is brought to the frequency to be tested, parked there for some time τ_{S} and read out afterwards. If there is no coupled TLS in the vicinity of the waiting position, the qubit will just decay exponentially yielding a more or less constant excitation probability independent of frequency. If the qubit is resonantly exchanging energy with a TLS, even if it is weakly coupled, the qubit's excitation probability will be less in comparison to the previous case most of the time (Fig. 4.6(b)). Therefore, this measurement using only one photon is

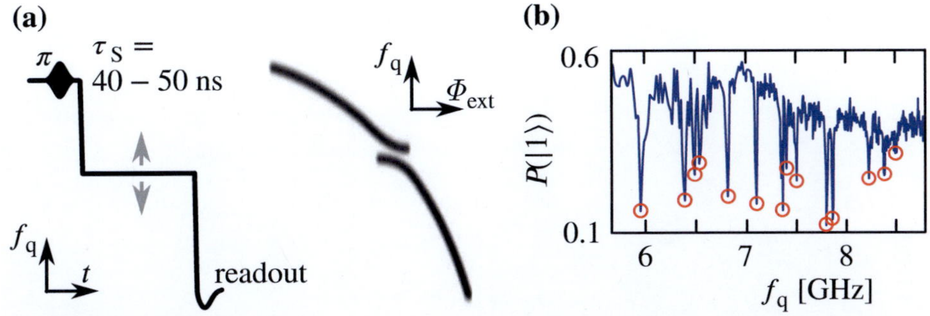

Figure 5.8: (a) The pulse sequence for the single photon spectroscopy (SPS) consists of a π-pulse on the qubit and a waiting time τ_S of about 45 ns. When the qubit hits a resonance of a TLS (indicated by the anti-crossing on the right) the measured qubit probability will be lower due to coherent energy exchange with the TLS. A resulting curve at one piezo bias is shown in **(b)**.

much more sensitive to TLSs [Sha+10]. Fig. 5.8(b) shows the qubit excitation probability as function of f_q at one piezo voltage with very sharp and deep dips representing TLS resonances. While still yielding better results, the resolution can be decreased, thus shortening the time needed to record a spectrum by a factor of three. It is also clear, that the described procedure hardly needs a calibration, except for the π-pulse. Due to the large microwave power required, it is not sensitive to small drifts of the qubit resonance. Only the recalculation from the flux bias to the frequency needs an additional measurement, which can also be done after taking the spectrum, if needed.

If a TLS is of special interest, the waiting time should be made equal to the swap time $\tau_S = 1/2v$, where v is the qubit-TLS coupling strength. For a spectroscopy measurement of all TLSs, τ_S should be chosen so that the spectrum reveals as many TLS traces as possible. To detect strongly ($\geq$ 30 MHz) as well as weakly ($\leq$ 10 MHz) coupled TLSs and to stay within the qubit's coherence times, we choose a τ_S in the range of 40 – 50 ns, usually 45 ns.

The SPS is generated by composing the corresponding curves like in Fig. 5.8(b). Usually, in our experiments, the qubit is excited at lower frequencies. This is recommended, because at larger bias flux the qubit potential has a stronger asymmetry which suppresses the excitation of the higher excited states. However, while tuning the qubit to higher frequencies and then back to lower energies for the readout, the qubit looses some of its population due to interactions with other TLSs "on the way", which can be seen as a gradient of

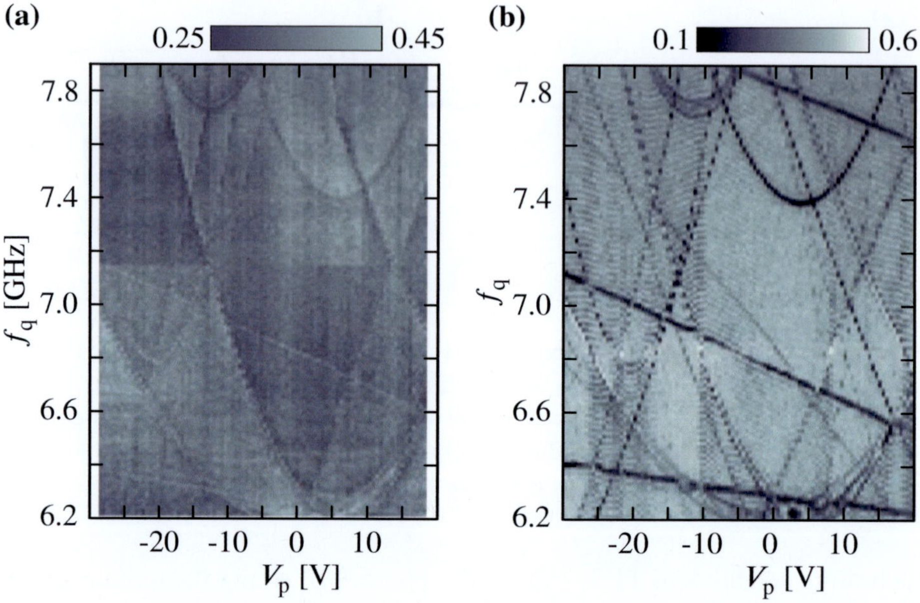

Figure 5.9: A comparison between the two spectra taken according to MPS **(a)** and SPS **(b)**. SPS is not only much more precise and has a much higher contrast, but it also only takes approximately one third of the time needed for MPS.

the maximal excitation probability as a function on the qubit frequency. This gradient, however, can be subtracted in data postprocessing.

Fig. 5.9 shows the comparison between MPS (a) and SPS (b). Already at first sight one can see the increased sharpness and better TLS trace visibility. While only $\approx$ 10 traces can be recognized on the MPS, the SPS reveals nearly twice as many. One can argue, of course, that the run of the MPS was badly calibrated, this can be in particular identified by the white-black double traces. Such structures appear when the expected qubit resonance differs slightly from the real one, e.g. due to flux drifts. In other words, the yellow curve in Fig. 5.5 is not lying directly on the qubit resonance but slightly off. Then, the qubit is excited less, and, more importantly, passing an anti-crossing the qubit population first increases and then decreases again to zero. This confirms the problem of calibration which is a very crucial task for taking the MPS. On the other hand, the SPS also yields some artifacts in its spectrum, which, however, do not affect the TLS visibility negatively. For example, choosing a fixed time τ_S for all TLSs has the disadvantage, that the traces of TLSs with a coupling strength to the qubit of $v \approx 1/\tau_S$ will appear as double curves with the real TLS

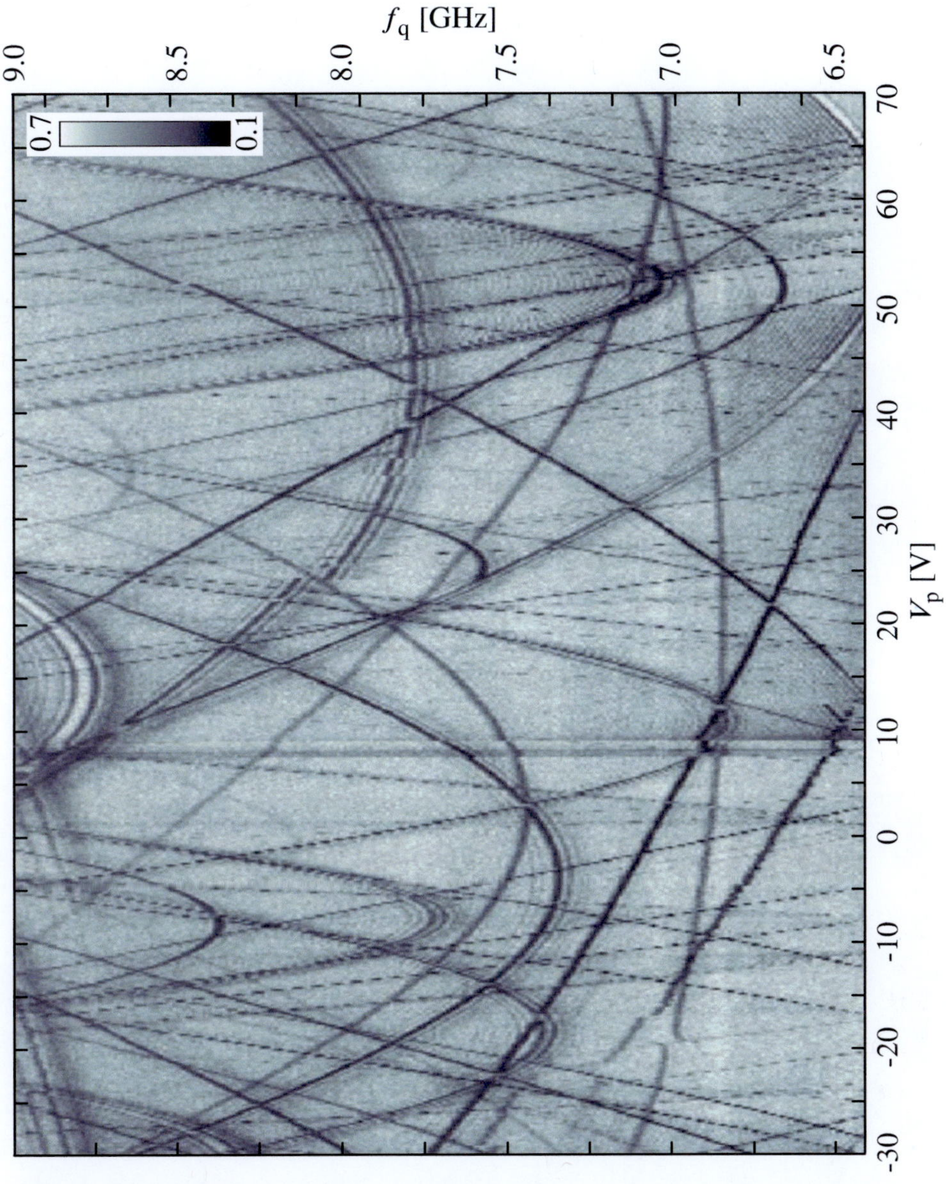

Figure 5.10: An SPS of the same qubit as in Fig. 5.11 taken in December, 2012. $P(|1\rangle)$ is encoded in color.

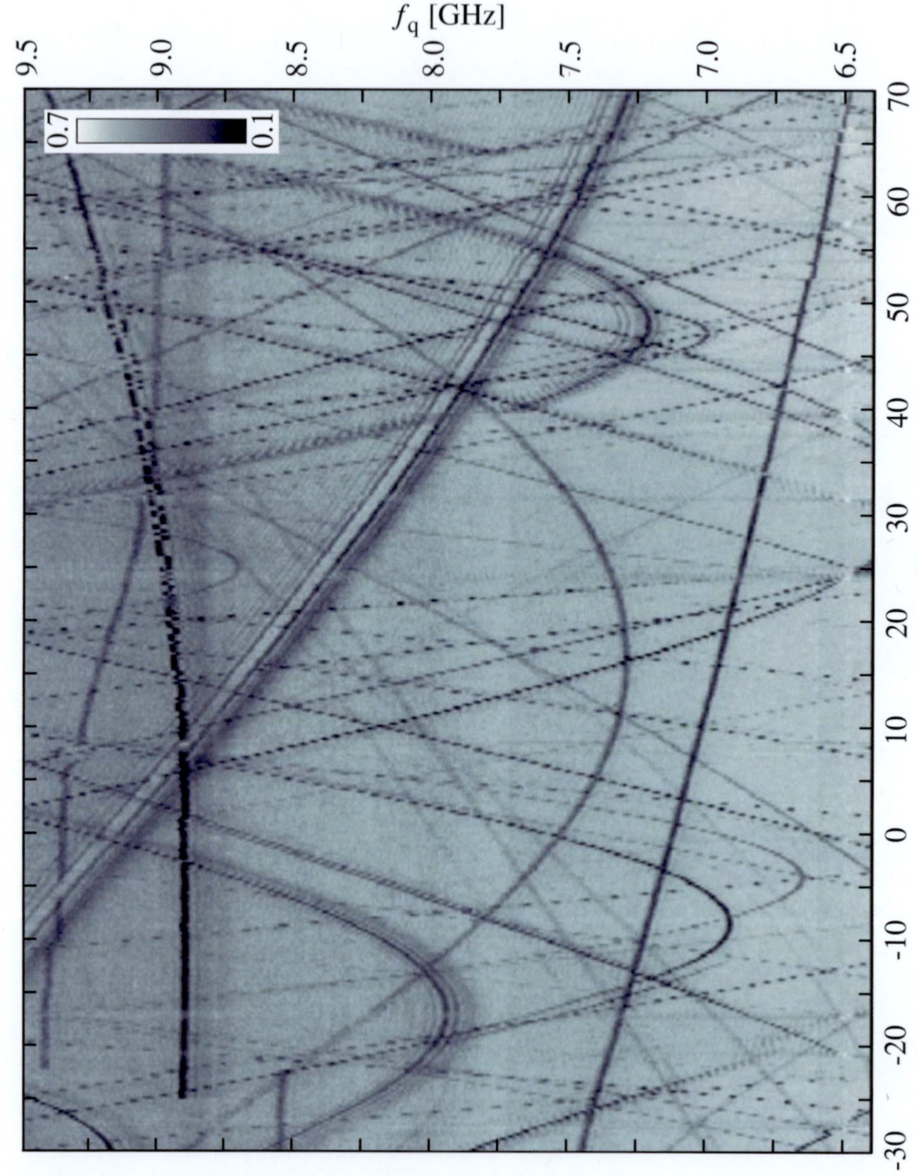

Figure 5.11: An SPS of the same qubit as in Fig. 5.10 taken in March, 2013. $P(|1\rangle)$ is encoded in color.

traces in-between. Being in resonance with such a TLS, the qubit gets its energy back after τ_S. Since the oscillation frequency between the qubit and the TLS increases with detuning, the minimal qubit excitation is measured at a frequency slightly different from the TLS resonance. This is also the explanation for the ''shadows'' visible around the TLS traces in general. Such structures can be recognized clearly in Fig. 5.9(b), e.g. the two hyperbolas at the very top or bottom of the plot and, in particular, in Figs. 5.10 and 5.11. Another point concerns the mentioned loss of the qubit energy. Exciting the qubit at lower energies, tuning it to higher frequencies and then back for the readout does not only decrease the qubit population, but also introduces interference effects as can be seen, e.g., in Fig. 5.10 between 40 and 70 V.

Figs. 5.10 and 5.11 were taken on the same qubit with a time difference of four months and a warm up to room temperature in-between. On both plots, many hyperbolic and linear traces can be recognized, but also several single abrupt jumps. In particular, in the upper part of Fig. 5.11, one TLS trace is permanently jumping between two frequencies. Another important observation is that the distribution of TLSs and their traces changed completely due to the warm up. Comparing SPSs of the same qubit but after a warm up to room temperature, one can find linear and hyperbolic traces with the same strain dependence but shifted in frequency and piezo voltage. However, many more temperature cycles and measurements are required to be able to identify individual traces arising from the same defects. To change the Al/AlO_x barrier completely one has to go close to the melting temperature, which the qubit would not survive[5]. Therefore, it is easily comprehensible that at room temperature a lot of atomic arrangements in the tunneling barrier of the JJ are conserved.

5.3 Two coupled TLSs

By chance, an SPS revealed a non-linear and also non-hyperbolic TLS trace (Fig. 5.12(a)). The form rather resembles a stretched ''S'' with an anti-crossing symmetrically on each side. We will refer to this TLS as TLS1. After further investigation, the first impression was confirmed that this shape originates from an additional TLS interacting with TLS1, called TLS2, which is, however, hardly visible on the spectrum.

[5]For example, crystalline sapphire has its melting point at about 2000 °C.

In principle, simple classical considerations of two interacting dipoles already explain the ''S''shape. Being in a stationary state, the dipole-dipole interaction (Eq. (5.34)) introduces an energy shift, preferring one configuration of two dipoles to another. This interaction also yields an energy exchange, i.e. oscillations of one dipole are transferred to the other, which in quantum mechanics results in an anti-crossing. In this sense, $g_\perp$ and $g_\parallel$, being the coupling strengths corresponding to the transversal energy exchange interaction and the longitudinal energy shift, respectively, can be defined and fitted. However, this model looks like being ''patched together'' and not all of the aspects resulting from the tunneling model are taken into account. In the following, the complete theory behind two interacting TLSs is presented, with one or the other surprising result.

5.3.1 The Hamiltonian and the "S"

The full tripartite Hamiltonian reads

$$H = H_1 + H_2 + H_\mathrm{q}^{(d)} + H_{1\mathrm{q}} + H_{2\mathrm{q}} + H_{12}. \tag{5.7}$$

H_i (i={1,2}) is the Hamiltonian of the ith TLS, which is not a priori diagonal,

$$H_i = \frac{1}{2}\varepsilon\sigma_z + \frac{1}{2}\Delta\sigma_x \tag{5.8}$$

with σ_j being the Pauli matrices, and ε and Δ the asymmetry and tunneling energy between the two wells, respectively. In contrast to the TLSs, the qubit states only exist in one potential well resembling a harmonic oscillator. Therefore, the qubit Hamiltonian is assumed to be diagonal initially, which is denoted by the superscript (d),

$$H_\mathrm{q}^{(d)} = \frac{1}{2}E_q\tau_z^{(d)} \tag{5.9}$$

with τ_j being the Pauli matrices referring to the qubit. Since the coupling between the qubit and the TLS is assumed to be of electrical nature [Mar+05], the corresponding qubit operator[6] is $\tau_x^{(d)}$, whereas the coupling of a TLS to its

[6] Another demonstrative explanation for the qubit coupling to E-fields to be $\tau_x^{(d)}$ is the following: In agreement with the double well potential, the two lowest states of a harmonic potential can also be defined as $|L, R\rangle = (|0\rangle \pm |1\rangle)/\sqrt{2}$. This corresponds to a particle located

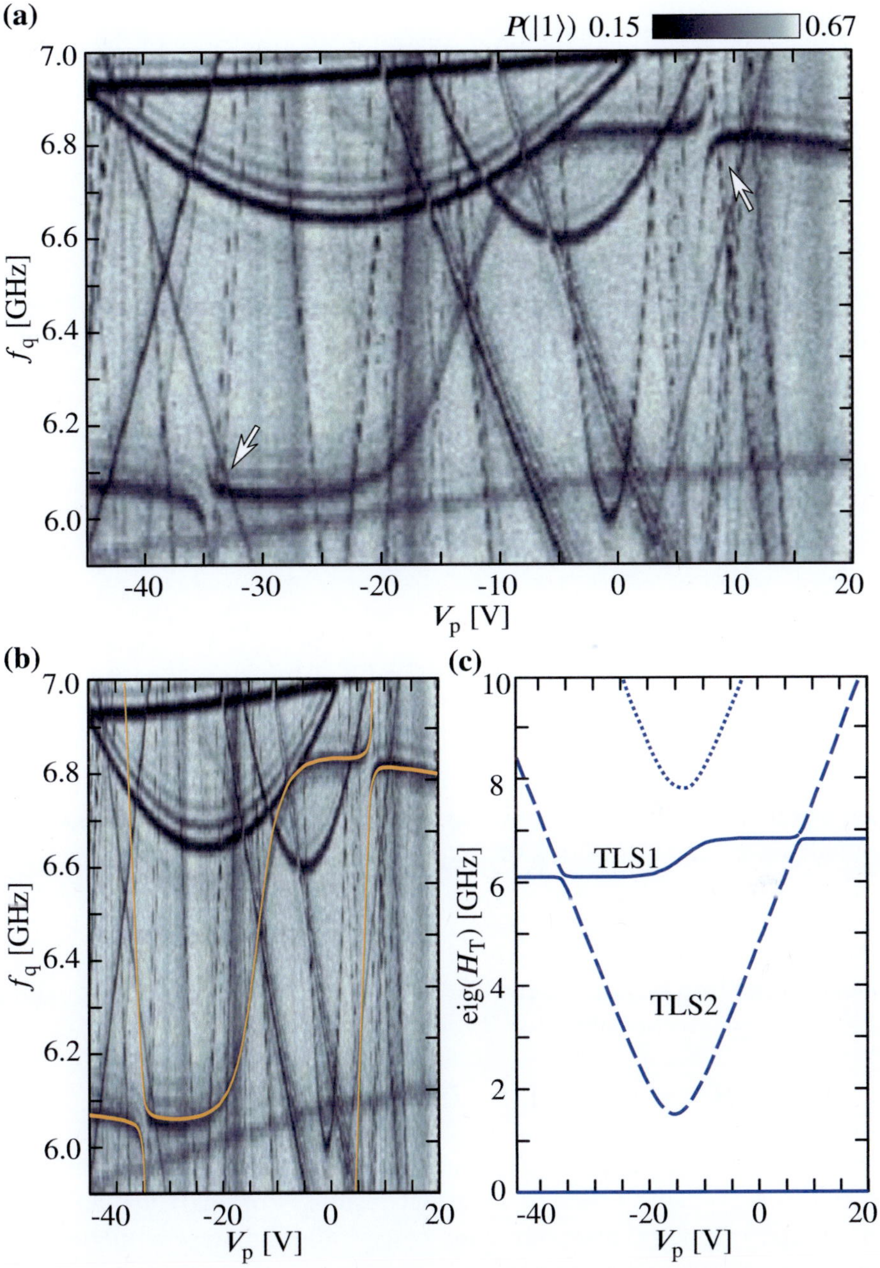

Figure 5.12: (a) SPS showing the region of interest. The S-shape and both anti-crossings (marked by white arrows) on each side of the "S" can be clearly seen. **(b)** The numerically calculated spectrum overlaid with the measurement shows excellent agreement. The full spectrum, including the TLS2 hyperbola (dashed) and the fully excited state (dotted), is plotted in **(c)**.

environment is established via the σ_z operator (modification of the asymmetry energy). Thus, the coupling terms are

$$H_{iq} = \frac{1}{2} v_i \sigma_{z,i} \tau_x^{(d)} \tag{5.10}$$

$$H_{12} = \frac{1}{2} g \sigma_{z,1} \sigma_{z,2}, \tag{5.11}$$

where the TLS1-TLS2 interaction is mediated via electric or elastic fields, or a combination of both. For a deeper analysis, it is more convenient to go into the eigenbasis of the corresponding subsystem. Therefore, we diagonalize the TLS Hamiltonians via the unitary transformation

$$U(\alpha_i) = \begin{pmatrix} \cos\frac{\alpha_i}{2} & \sin\frac{\alpha_i}{2} \\ -\sin\frac{\alpha_i}{2} & \cos\frac{\alpha_i}{2} \end{pmatrix} \tag{5.12}$$

with $\tan \alpha_i = -\Delta_i/\varepsilon_i$ (Eqs. (2.8) and (2.9))[7]. The choice of $U(\alpha_i)$ is so that

$$H_i^{(d)} = U(\alpha_i)^\dagger H_i U(\alpha_i) = \frac{1}{2} E_i \sigma_{z,i}^{(d)}. \tag{5.13}$$

It is important to also mark the corresponding operators with the superscript (d). Although having the same mathematical representation, $\sigma_j^{(d)}$ are referring to the eigenbasis of the TLSs, whereas σ_j operate on the physical space. The diagonalization also transforms the $\sigma_{z,i}$ operators,

$$\sigma_{z,i} \to U(\alpha_i)^\dagger \sigma_{z,i} U(\alpha_i) = \sin \alpha_i \sigma_{x,i}^{(d)} + \cos \alpha_i \sigma_{z,i}^{(d)}, \tag{5.14}$$

and so the coupling Hamiltonians. The full Hamiltonian in Eq. (5.7), expressed in the eigenbasis of the subsystems, reads

$$H^{(d)} = H_1^{(d)} + H_2^{(d)} + H_q^{(d)} + H_{1q}^{(d)} + H_{2q}^{(d)} + H_{12}^{(d)}. \tag{5.15}$$

For numerical calculations and simulations (Sec. 2.2), this one or the Hamiltonian in Eq. (5.7) can be used to get exact results. For analytical analysis and demonstrative representation, however, some important approximations can be

rather on the left or the right side (as the right and the left well) so that the coupling to the E-fields is mediated by the operator $\tau_z^{(d)}$. After diagonalization, we get the eigenstates $|0\rangle$ and $|1\rangle$ while $\tau_z^{(d)}$ transforms to $\tau_x^{(d)}$.

[7] It is assumed that Δ_i is real.

applied to simplify the problem significantly. The three coupling Hamiltonians consist of two (Eq. (5.10)) or even four terms (Eq. (5.11)), e.g.

$$H_{12}^{(d)} = \frac{1}{2}g(\cos\alpha_1 \cos\alpha_2\, \sigma_{z,1}^{(d)} \sigma_{z,2}^{(d)} + \sin\alpha_1 \sin\alpha_2\, \sigma_{x,1}^{(d)} \sigma_{x,2}^{(d)} + \cos\alpha_1 \sin\alpha_2\, \sigma_{z,1}^{(d)} \sigma_{x,2}^{(d)} + \sin\alpha_1 \cos\alpha_2\, \sigma_{x,1}^{(d)} \sigma_{z,2}^{(d)}). \tag{5.16}$$

The ''xx''- and ''zz''-terms are easy to handle since these common couplings introduce an anti-crossing for resonant systems and a constant energy shift, respectively, as explained in the introduction to this section. The idea is to neglect all ''xz''- and ''zx''-terms. For example, the matrix representation of the $\sigma_{z,1}^{(d)}\sigma_{x,2}^{(d)}$-operator is

$$\sigma_{z,1}^{(d)}\sigma_{x,2}^{(d)} = \begin{pmatrix} 0 & 1 & 0 & 0 \\ 1 & 0 & 0 & 0 \\ 0 & 0 & 0 & -1 \\ 0 & 0 & -1 & 0 \end{pmatrix}, \text{basis} = \begin{bmatrix} |ee\rangle \\ |eg\rangle \\ |ge\rangle \\ |gg\rangle \end{bmatrix}, \tag{5.17}$$

where $|eg\rangle$, e.g., means that TLS1 is excited and TLS2 is in its ground state. This corresponds to a modification of the eigenstates of TLS2 as a function of the population of TLS1 without introducing an anti-crossing. The only remaining question is whether the prefactors of the coupling terms are small enough to be neglected. As it is shown in the following, the results won't change qualitatively. For the qubit-TLS coupling this is obvious. While the qubit energies E_q are between 5 and 10 GHz, the coupling strength to the TLSs v is not larger than $\approx$ 50 MHz according to our observations. The $\tau_x^{(d)}\sigma_{z,i}^{(d)}$-term slightly changes the qubit eigenenergy by a maximum of $\sqrt{(7\,\mathrm{GHz})^2 + (50\,\mathrm{MHz})^2} - 7\,\mathrm{GHz} \approx 0.2\,\mathrm{MHz}$. In the case of TLS1-TLS2 coupling, one has to pay more attention. In our case, the interaction strength is $g = -872\,\mathrm{MHz}$. As can be seen in Fig. 5.12, the TLS1 energy hardly changes with strain so that α_1 remains constant with the estimated value of 60° (Tab. 5.1, Sec. 5.3.2). The largest of the two coupling strengths to be neglected is $\propto \sigma_{x,1}^{(d)}\sigma_{z,2}^{(d)}$ with the prefactor $|g \sin\alpha_1 \cos\alpha_2| \leq 755\,\mathrm{MHz}$ ($\alpha_2 = 0$). The maximal induced energy shift is then $\sqrt{(7\,\mathrm{GHz})^2 + (755\,\mathrm{MHz})^2} - 7\,\mathrm{GHz} \approx 40\,\mathrm{MHz}$. Although not very small, this frequency change is still an order of magnitude smaller than the longitudinal energy shift $\propto \sigma_{z,1}^{(d)}\sigma_{z,2}^{(d)}$ of $|g\cos\alpha_1 \cos\alpha_2| \leq 428\,\mathrm{MHz}$ and approximately one forth of the size of the TLS1-TLS2 anti-crossing. Therefore, only the perpendicular and longitudinal coupling strengths are important for qualitative agreement between theory and experiment:

$$v_{i,\perp} = v_i \sin \alpha_i, \ g_\perp = g \sin \alpha_1 \sin \alpha_2, \ g_\parallel = g \cos \alpha_1 \cos \alpha_2. \tag{5.18}$$

Thus, the significant part of the full Hamiltonian, (Eq. 5.15), is

$$H_{\text{sig}}^{(d)} = \frac{1}{2}(E_{\text{q}}\tau_z^{(d)} + E_1\sigma_{z,1}^{(d)} + E_2\sigma_{z,2}^{(d)} + g_\parallel \sigma_{z,1}^{(d)}\sigma_{z,2}^{(d)} + g_\perp \sigma_{x,1}^{(d)}\sigma_{x,2}^{(d)}$$
$$+ v_{1,\perp}\sigma_{x,1}^{(d)}\tau_x^{(d)} + v_{2,\perp}\sigma_{x,2}^{(d)}\tau_x^{(d)}). \tag{5.19}$$

As mentioned before, in the numerical simulations the full Hamiltonian is taken into account. Based on $H_{\text{sig}}^{(d)}$, we can make the following link to the ideas in the beginning of this section, where two oscillating and stationary dipoles were considered classically. The expectation value of the location of the atom, and so its charge, is $\langle \sigma_z \rangle = \cos \alpha$. In other words, $g_\parallel = g \left\langle \sigma_{z,1} \right\rangle \left\langle \sigma_{z,2} \right\rangle$ corresponds to the classical energy shift due to dipole-dipole interaction. On the other hand, a quantum mechanical evolution of the wave function corresponding to oscillations of the expectation value of the charge position can only be achieved by a superposition of the two TLS states. Therefore, we have to know the expectation value of the $|\pm\rangle$ states, or $\langle \sigma_x \rangle = \sin \alpha$, and so $g_\perp = g \left\langle \sigma_{x,1} \right\rangle \left\langle \sigma_{x,2} \right\rangle$.

5.3.2 Characterizing the TLS1-TLS2 system

Considering solely the interacting TLS1-TLS2 system, we define the Hamiltonian H_{T} as

$$H_{\text{T}} = H_1 + H_2 + H_{12}. \tag{5.20}$$

This Hamiltonian has seven parameters: three for each TLS (tunneling energy Δ, deformation potential γ and voltage offset for which the asymmetry energy ε is zero) and the coupling strength g. In principle, all these parameters can be determined by fitting the S-shape (see below). However, fitting a single curve with so many degrees of freedom is quite a dirty solution. A much better idea, as will be explained in the following, is to also measure the energy of the full excited TLS1-TLS2 system at the strain where both systems are in resonance.

Measuring the energy of the full excited state

We considered several solutions. Obviously, we have to bring two excitations of different energies into the system (Fig. 5.13(a)). A straight forward way would

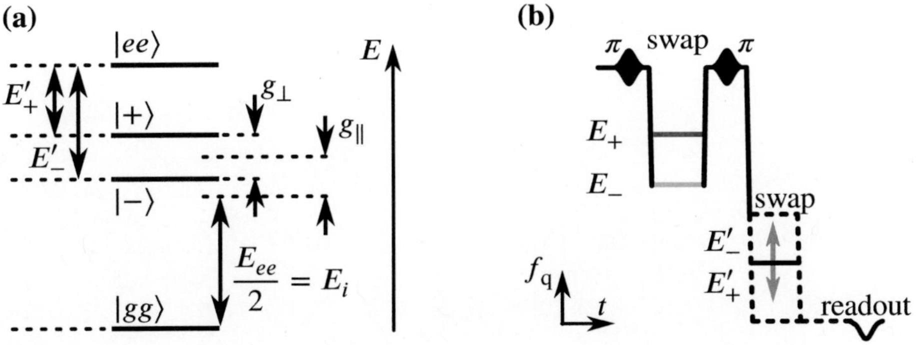

Figure 5.13: (a) The level structure at the right TLS1-TLS2 anti-crossing is shown. Measuring the frequencies $E'_{\pm} = E_{ee} - E_{\pm}$ allows to determine the energies of the unperturbed TLSs E_1 and E_2 and the coupling strengths $g_{\perp}$ and $g_{\parallel}$ directly. **(b)** The pulse sequence to measure E_{ee} is shown schematically. After exciting the qubit, its excitation is swapped to one of the $|\pm\rangle$ states. Afterwards, the SPS protocol should reveal the energies $E'_{\pm}$.

be to use two microwave sources, which, however, complicates the experimental setup, and the usage of additional microwave components introduces new sources of noise and reflections. Another possibility would be to fully excite the TLS1-TLS2 system directly by a resonant microwave pulse or via a two-photon-absorption [Bus+10]. This idea was simulated using our program and it turns out that the coupling strength is too low to achieve measurable results. Much more promising are the following two pulse sequences using the qubit to excite the TLSs. For example, one can populate the qubit into the state $|2\rangle$ and then tune the $|1\rangle \leftrightarrow |2\rangle$ transition into resonance with one of the $|\pm\rangle$ branches (TLS1 and TLS2 are in resonance). After the swap, the now singly excited qubit is used to find the missing energy to fully excite the TLS1-TLS2 system. Another idea is to singly excite the qubit, swap its population to one of the two branches, and subsequently perform the SPS protocol (Sec. 5.2.2) in the frequency range of interest. After analyzing the simulation results, we decided to concentrate on the latter sequence.

Fig. 5.13(a) shows the level structure of the TLS1-TLS2 anti-crossing on the right of the ‘‘S’’ where the energies of the $|\pm\rangle$ states are shifted to higher frequencies. The pulse sequence to find the missing energies $E'_{\pm} = E_{ee} - E_{\pm}$ which bring the TLS1-TLS2 system into to the full excited state $|ee\rangle$ is displayed in Fig. 5.13(b). After exciting the qubit with a π-pulse, its excitation is swapped either to the state $|+\rangle$ or $|-\rangle$, which takes $\approx 50\,\text{ns}$. Subsequently, the qubit is

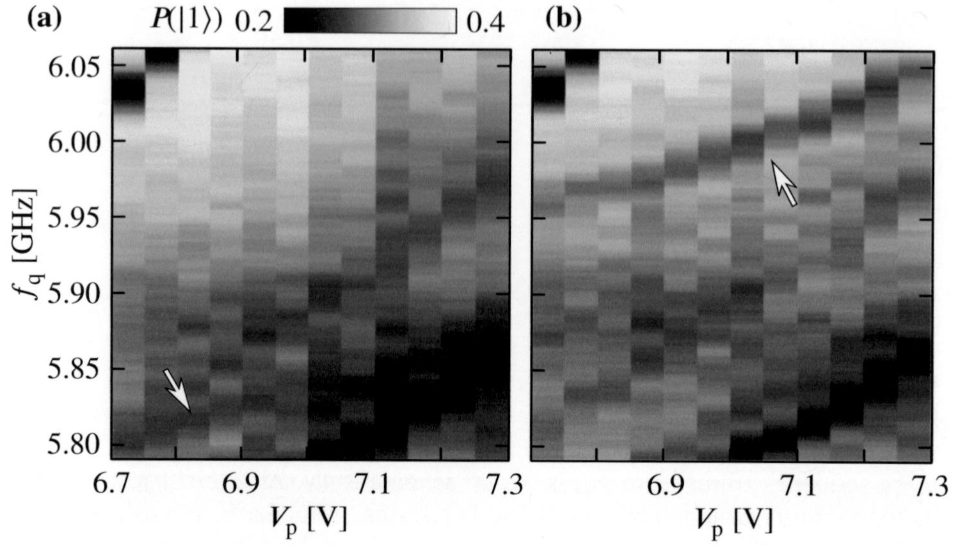

Figure 5.14: The results of the pulse sequence from Fig. 5.13(b). The arrows on the plots show the additional traces appearing if swapping the qubit excitation first to the upper **(a)** (E'_+) or the lower **(b)** (E'_-) branch.

excited again and then tuned to the frequency of interest, where it resides again for the swap pulse duration. The two SPSs taken in such a way, one for each of the two branches, then show two additional TLS traces (Fig. 5.14). To show the result more clearly, one plot is subtracted from the other (Fig. 5.15(b)). On the one hand this procedure completely cancels the background structure, on the other hand the two traces of interest appear in the new plot as dips (red, E'_+) and as peaks (green, E'_-).

Determination of the parameters and simulation results

As mentioned in the beginning of this section, in principle, all seven parameters of the Hamiltonian in Eq. (5.20) can be estimated by fitting the S-curve from Fig. 5.12 due to following reasons. The three parameters of the TLS2 hyperbola are fixed by the two anti-crossings, and so the angle α_2 at the TSL1-TSL2 resonance is known. As can be seen from Fig. 5.13(a), the amplitude of the "S" is $2g_{\parallel}$ and the splitting size is $g_{\perp}$, which are directly related to the bare TLS1-TLS2 coupling strength g and α_1 (Eq. 5.18). Finally, α_1 determines the decomposition of the TLS1 frequency into the tunneling and asymmetry energy, which in turn yields the deformation potential of TLS1.

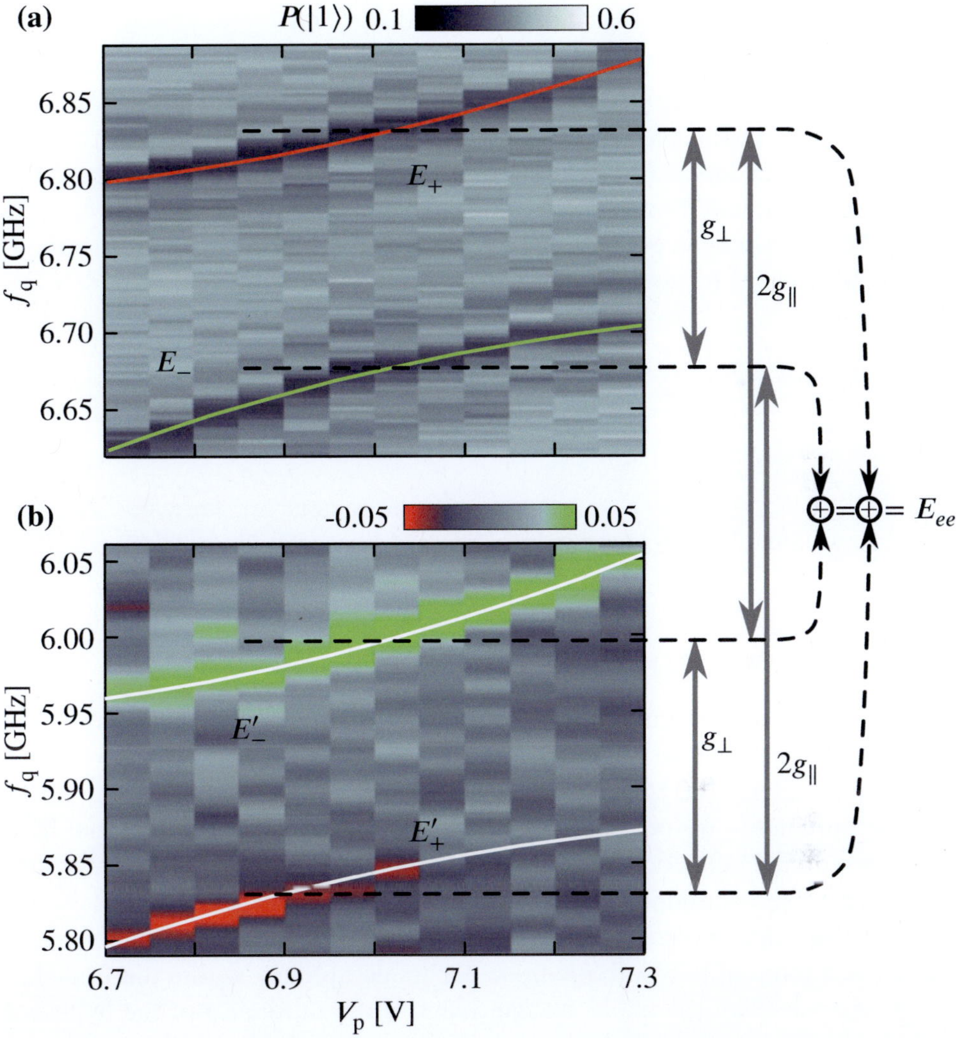

Figure 5.15: Some relevant parameters can be estimated directly from the right TLS1-TLS2 anti-crossing **(a)** and the corresponding energies to fully excite the TLS1-TLS2 system **(b)**. The size of the anti-crossings equals $g_\perp$ while the energetical difference between them corresponds to $2g_\parallel$. Adding up the energies as shown by black arrows results in E_{ee} which equals twice the unperturbed TLS energies E_i.

Although all parameters seem to be responsible for different features of the ‘‘S’’, it is very difficult to extract them ‘‘one by one’’. However, a step into this direction is the knowledge of the energies of the complete TLS1-TLS2 system at their resonance. Instead of one, the parameters are now determined in three independent steps. First, only the three parameters of the TLS2 hyperbola are fitted from the spectrum. Then, as shown in Fig. 5.15, at the TLS1-TLS2 resonance $g_\perp$, $g_\|$, and the unperturbed TLS energies E_1 and E_2 can be determined by using the formulas

$$\begin{aligned} 2E_1 &= 2E_2 = E_+ + E'_+ = E_- + E'_- = E_{ee} \\ g_\perp &= E_+ - E_- = E'_- - E'_+ \\ 2g_\| &= E_+ - E'_- = E_- - E'_+, \end{aligned} \tag{5.21}$$

with $E'_\pm = E_{ee} - E_\pm$. Now, almost all parameters are known. The only remaining value, γ_1, can be easily estimated from the slope of TLS1 in the S-curve.

Unexpected frequency jump

Unfortunately, approximately two weeks passed between the measurement of the S-shape and the additional experiment; and the TLS1 energy changed by approximately 100 MHz towards lower frequencies. In general, during our measurements we occasionally observe abrupt jumps of individual TLS’ resonance frequencies on a time-scale of days. The jumps are more frequent directly after the cool down and their occurrence decreases very strongly with time so that after several weeks of operation the TLS frequencies are mostly stable. If a jump of a hyperbolic trace occurs completely within the frequency range accessible by the qubit, we can identify the minimum of the hyperbola as being shifted in both, the piezo voltage as well as the frequency, suggesting that the asymmetry and tunneling energy have changed. However, we never detected a change of the deformation potential $\gamma \propto \delta\varepsilon_1/\delta V_\mathrm{p}$, i.e. a modified slope or curvature of the traces. For a full characterization of the system we further have to assume that Δ_2 did not change appreciably in comparison to ε_2 at the TLS1-TLS2 anti-crossing. We can prove these assumptions by comparing the numerically calculated spectrum and the left TLS1-TLS2 anti-crossing after the jump. Taking the parameters Δ_2, γ_2 and γ_1 from the ‘‘S’’, and all other values from the additional experiment (Tab. 5.1), a good agreement can be found (Fig. 5.16).

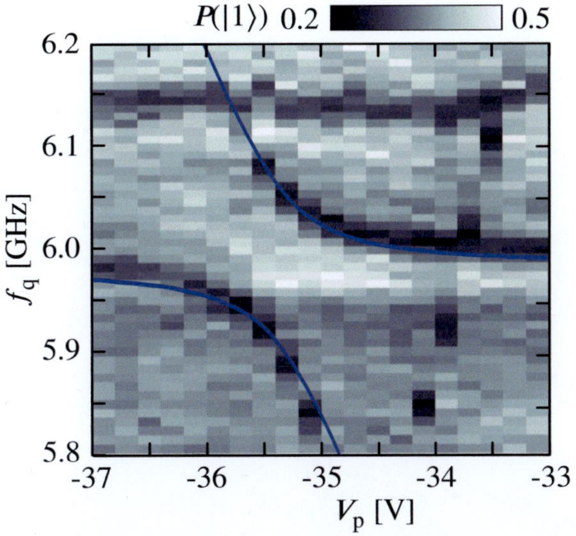

Figure 5.16: The measured left TLS1-TLS2 anti-crossing after the jump and the simulated spectrum (with the values in Tab. 5.1) show good agreement. This confirms that the TLS2 energies remain mostly unchanged.

Nevertheless, one can try to find out which of the two parameters, ε_1 or Δ_1, or their combination, is responsible for the frequency jump of TLS1 towards lower energies. Increasing only ε_1 yields a larger $g_{\parallel}$ (Eq. (5.18)), while adapting the TLS1 energy using only Δ_1 reduces the "S"-amplitude. It turns out that the spectrum of Fig. 5.12 can be very accurately reproduced by just increasing Δ_1 from 5.47 GHz to 5.55 GHz and slightly shifting the TLS2 hyperbola to the right by changing the voltage of the symmetry point of TLS2 from -14.05 V to -13.55 V (Fig. 5.12(b)).

Finally, Fig. 5.17 shows the simulation results of the pulse sequences presented in this section. The SPS protocol yielding the full spectrum is shown in Fig. 5.17(a), where the 100 MHz jump is taken into account. The simulation of the additional experiment measuring the energies of the TLS1-TLS2 level structure (Fig. 5.15) is presented in Fig. 5.17(b) and (c).

5.3.3 The qubit-TLS2 coupling

Experiments

By looking very closely on the right anti-crossing of the "S", it seems that the upper branch vanishes faster than the lower, which was confirmed by

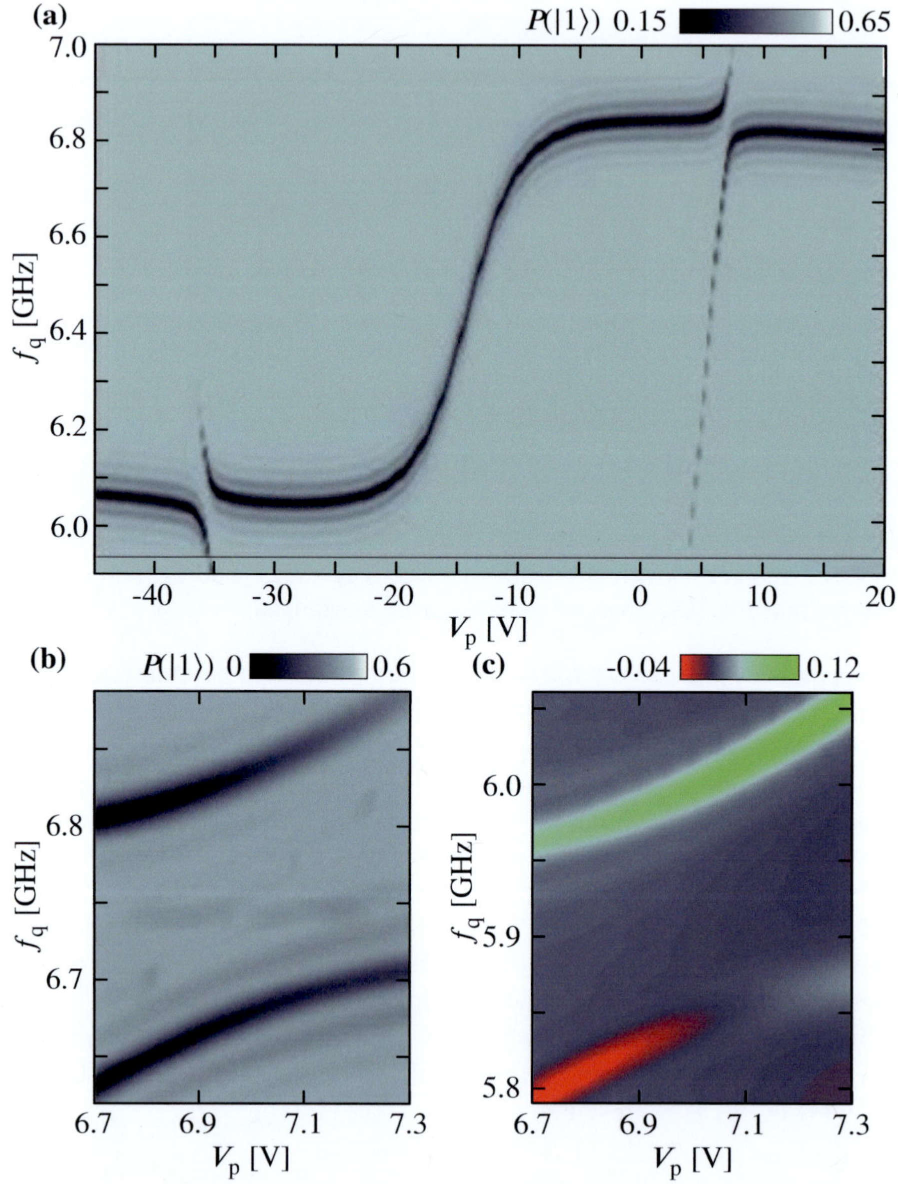

Figure 5.17: (a) The simulation of the full spectrum according to the SPS protocol perfectly reproduces the "S" with the anti-crossings. The 100 MHz jump of TLS1 is taken into account. **(b)** and **(c)** show the simulated results being in excellent agreement with the additional experiment characterizing the TLS1-TLS2 level structure at the right anti-crossing.

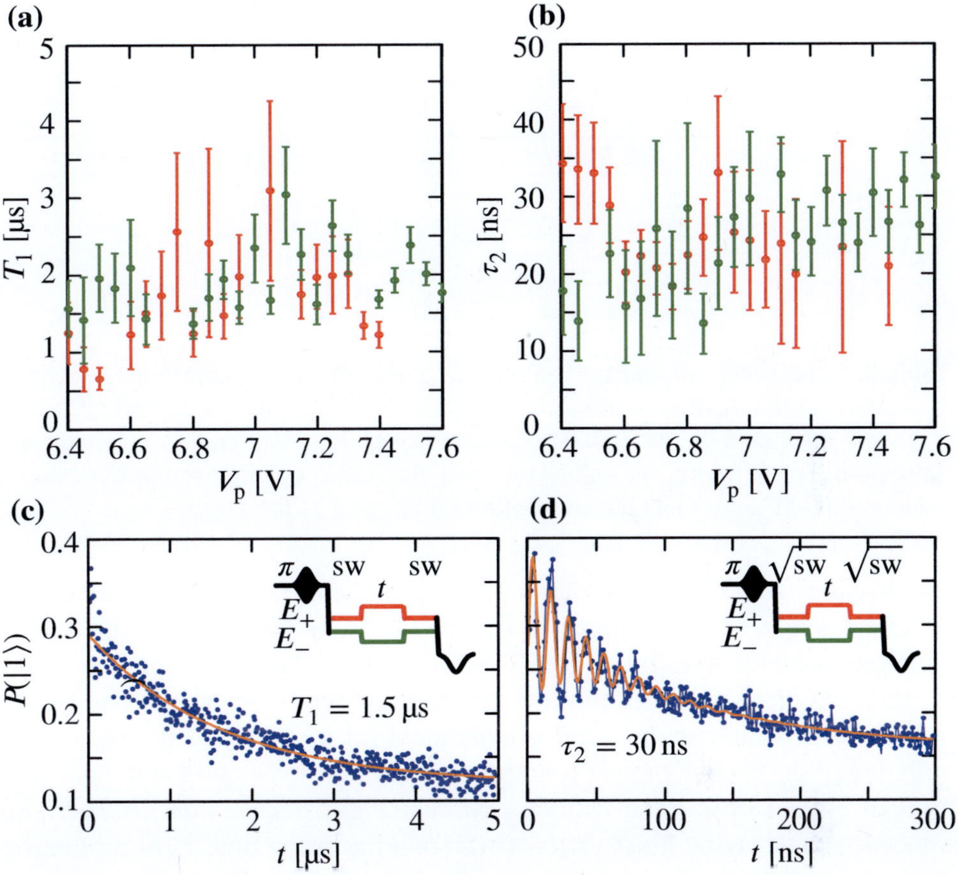

Figure 5.18: (a) The relaxation times T_1 of the upper (red) and the lower (green) branches were measured in the vicinity of the right TLS1-TLS2 anti-crossing. The corresponding pulse sequence ("sw" is a swap) and an example of the decay is shown in **(c)**. **(b)** We tried to estimate the dephasing time of the two branches by entangling the qubit with the $|\pm\rangle$ states (see text). Since the decay time of the beating suffers from qubit decoherence, it is denoted as τ_2. The pulse sequence and an example of the oscillations is shown in **(d)**. By simulating this pulse sequence, T_2 could be estimated.

	Δ_i	$\varepsilon_i(V_\text{p})$ @ $\varepsilon_2 = 0$	ξ_i @ $E_1 = E_2$	$\lvert v_{i,\perp} \rvert$@ $E_1 = E_2$	$\lvert v \rvert$
TLS1	5.47 GHz	3.18 GHz – 4 MHz/V	60°	15.4 MHz	18.7 MHz
TLS2	1.3 GHz	295 MHz/V	12°	3 MHz	14.3 MHz

$g_\parallel = -428$ MHz
$\lvert g_\perp \rvert = 155$ MHz
$g = -872$ MHz

Table 5.1: This table summarizes the relevant parameters. The asymmetry energy ε_i is given at the TLS2 symmetry point ($V_\text{p} = -14.05$ V). The parameters ξ_i and $v_{i,\perp}$ were estimated at the TLS1-TLS2 resonance at $V_\text{p} = 7$ V and so v_i could be calculated. Furthermore, the values for the TLS1-TLS2 coupling g and at the same piezo voltage ($V_\text{p} = 7$ V) its decomposition into $g_\parallel$ and $g_\perp$ are given.

more accurate measurements (data not shown). One possible explanation for this effect might be a difference in coherence times between both branches. Therefore, we tried to estimate the T_1 and T_2 times. The relaxation time could be measured directly by exciting the qubit, swapping its energy to one of the two branches, and after the qubit was detuned for some time t the remaining excitation of the excited branch was swapped back onto the qubit and measured (inset in Fig 5.18(c)). The resulting curves (Fig. 5.18(c)) were fitted by an exponentially decaying function [Nee+08] and the decay time T_1 as a function of the piezo voltage is plotted in Fig. 5.18(a). Due to the large relaxation times for both TLSs ($T_1 \approx 2$ µs) and a sufficiently large detuning of the qubit during the waiting time, any qubit influence can be excluded. The results show that neither a T_1 dependence on strain nor on branch can be recognized. The measurement of the dephasing time T_2 was a more challenging task. T_2 was known to be quite small so that the standard Ramsey sequence would not work. We decided to use the qubit as an auxiliary system and measured the oscillations resulting from entangling the qubit with one of the two branches by a $\sqrt{\text{sw}}$-gate, detuning the qubit from the branch for some time t, and performing an additional $\sqrt{\text{sw}}$-gate. The qubit excitation probability as a function of t (Fig. 5.18(d)) was fitted by damped oscillating curves and the decay time τ_2 was extracted. Again, no significant change of τ_2 can be seen. Due to the complex time evolution involving three quantum systems, we completely relied on simulation to extract T_2. By taking the qubit coherence times of 110 ns (relaxation) and

105 ns (dephasing), a τ_2 of 25 ns could be achieved by setting $T_2 = 70$ ns for both TLSs. In summary, the coherence times do not explain the asymmetry in visibility of the two branches, but even rather suggest that they should be approximately the same for both TLSs.

However, we found an asymmetry in the coupling strength between the qubit and the $|\pm\rangle$ states. By exciting the qubit and parking it in resonance with one or the other branch for some time (Fig. 5.19(a)), the qubit population shows oscillations (Figs. 5.19(c) and (d)), the frequency f_{osc} of which equals the coupling strength, as a function of the piezo voltage V_{p} (Fig. 5.19(b)). Indeed, while increasing V_{p}, the coupling between the qubit and the upper branch decreases faster than that with the lower branch. As can be seen in Fig. 5.19(a), the measured coupling strengths for each branch do not cross at $V_{\mathrm{p}} = 7$ V, where it would be expected if $\upsilon_{2,\perp} = 0$. If, however, $\upsilon_{2,\perp} \neq 0$, the shift of the coupling strength, and also the difference in visibility, can be explained.

Theory

This can be seen very easily by the following considerations. Under the condition that there is no TLS-qubit coupling ($\upsilon_{i,\perp} = 0$) but only TLS1-TLS2 interaction of strength $\tilde{g}$ (in principle, $\tilde{g} = g_\perp$), the 3×3 matrix of the single excitation space of the TLS1-qubit-TLS2 system has in the basis of diagonalized subsystems the form

$$\frac{1}{2}\begin{pmatrix} \Delta E & 0 & \tilde{g} \\ 0 & 0 & 0 \\ \tilde{g} & 0 & -\Delta E \end{pmatrix}, \text{ basis} = \begin{bmatrix} \text{TLS1} \\ \text{qubit} \\ \text{TLS2} \end{bmatrix}, \tag{5.22}$$

where $\Delta E = E_1 - E_2$ is the detuning between the two TLSs, the relative qubit energy does not matter. By diagonalizing this matrix with its eigenvectors

$$B = \begin{pmatrix} \cos\frac{\alpha}{2} & 0 & \sin\frac{\alpha}{2} \\ 0 & 1 & 0 \\ -\sin\frac{\alpha}{2} & 0 & \cos\frac{\alpha}{2} \end{pmatrix} \tag{5.23}$$

with $\tan\alpha = -\tilde{g}/\Delta E$ we go into the $\{|+\rangle\, |\mathrm{q}\rangle\, |-\rangle\}$ basis, where the $|\pm\rangle$ states are superpositions of the TLS1 and TLS2 eigenstates.

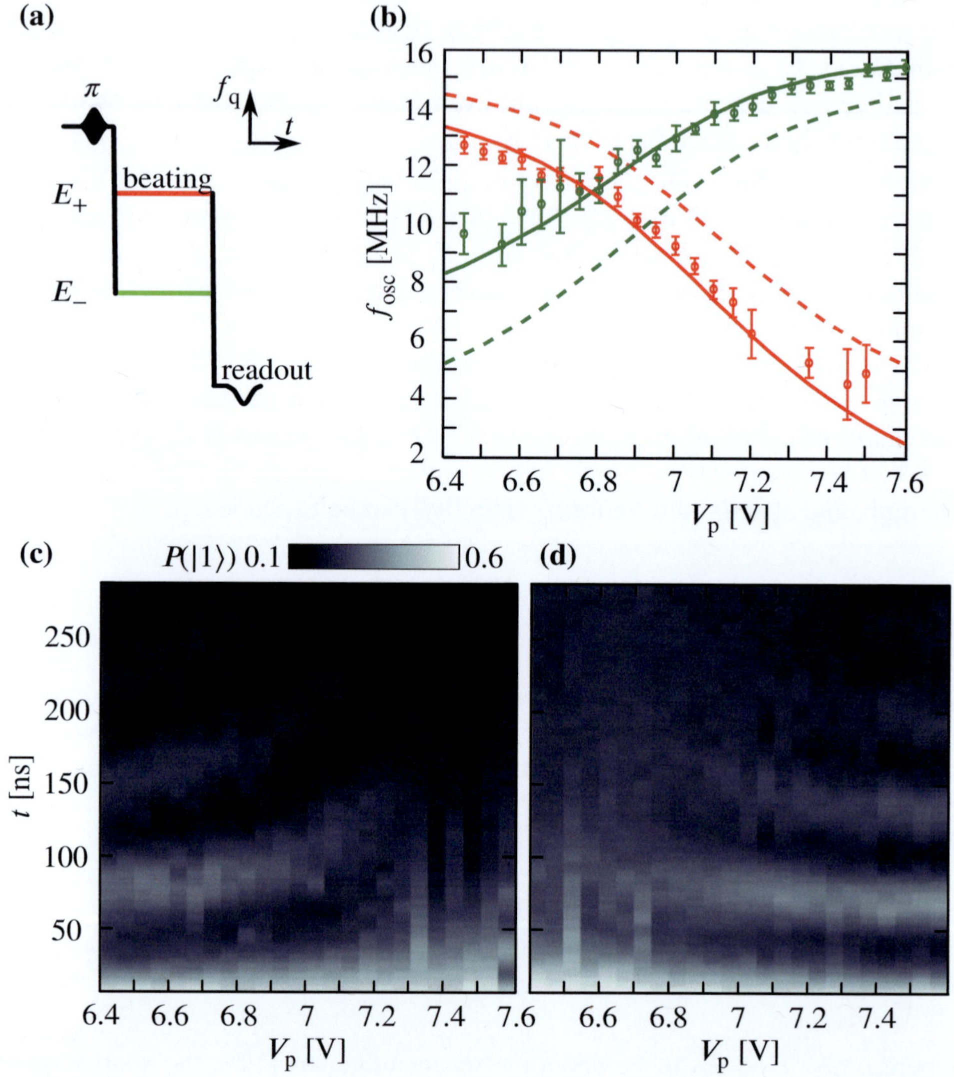

Figure 5.19: (a) To measure the coupling strength between the qubit and the $|\pm\rangle$ states, the qubit is excited and tuned into resonance with one of the two branches. The beating, as a function of strain, is visible for the upper and the lower branch on **(c)** and **(d)**, respectively. By fitting the time domain curves with damped oscillations, the beating frequencies f_{osc} can be extracted **(b)**. Dashed lines correspond to the case when $v_2 = 0$. The obvious shift of the expected frequencies can be explained by introducing a non-zero qubit-TLS2 coupling strength (see text).

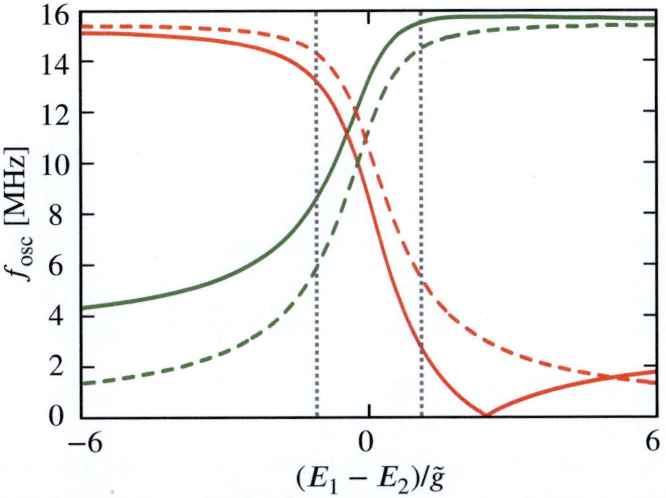

Figure 5.20: The calculated coupling strength in terms of $\tilde{g}$ between the qubit and both branches with $\upsilon_{2,\perp} = 0$ (dashed) and $\upsilon_{2,\perp} = 3\,\mathrm{MHz}$ (solid) ($\upsilon_{1,\perp} = 15.4\,\mathrm{MHz}$). The region between the dotted lines corresponds to the measured range assuming that $\tilde{g} = g_\perp$. As can be recognized, at some TLS1-TLS2 detuning the qubit-upper branch coupling equals zero.

If now coupling TLS1 and TLS2 to the qubit and looking at the situation from the $\{|+\rangle\,;|q\rangle\,;|-\rangle\}$ basis,

$$B^\dagger\,\frac{1}{2}\begin{pmatrix}\Delta E & \upsilon_{1,\perp} & \tilde{g}\\ \upsilon_{1,\perp} & 0 & \upsilon_{2,\perp}\\ g & \upsilon_{2,\perp} & -\Delta E\end{pmatrix}B = \frac{1}{2}\begin{pmatrix}\sqrt{\Delta E^2+\tilde{g}^2} & \upsilon_{+,\perp} & 0\\ \upsilon_{+,\perp} & 0 & \upsilon_{-,\perp}\\ 0 & \upsilon_{-,\perp} & -\sqrt{\Delta E^2+\tilde{g}^2}\end{pmatrix}, \qquad (5.24)$$

with

$$\upsilon_{\pm,\perp} = \upsilon_{1,\perp}\cos\frac{\alpha}{2} \pm \upsilon_{2,\perp}\sin\frac{\alpha}{2} \qquad (5.25)$$

being the coupling strength between the qubit and the upper and the lower branch, respectively (Fig. 5.20). Please note that with the definition of the angle α, at the TLS1-TLS2 resonance ($\Delta E = 0$) the upper branch is coupled to the qubit less strongly than the lower one, so that the oscillation frequencies between the qubit and the two branches are $f_{osc} = (\upsilon_{1,\perp} \mp \upsilon_{2,\perp})/\sqrt{2}$. The fit

(Fig. 5.19(b)) yields $\upsilon_{1,\perp} = 15.4\,\text{MHz}$ and $\upsilon_{2,\perp} = 3\,\text{MHz}$ for the transversal coupling strengths. Recalculating the bare coupling strengths from this, υ_i, using the known angles α_i yields a surprising result (Tab. 5.1). The reason for the very small value of $\upsilon_{2,\perp}$ is not a small dipole moment or its orientation, as one might think on first sight. On the contrary, υ_2 is even comparable to υ_1. The point is that at the TLS1-TLS2 anti-crossing TLS2 is already very asymmetric yielding a small α_2 and therefore a hardly measurable $\upsilon_{2,\perp}$.

5.3.4 An attempt to estimate the TLS1-TLS2 distance

On the one hand, the evaluation of the distance between TLS1 and TLS2 is based on the equation given in [Mar+05] relating the qubit-TLS coupling strength to its dipole moment. This equation will be derived in the following. On the other hand, the estimated dipole sizes of both TLSs and some particular configurations will be used to get an idea of the TLS1-TLS2 distance via the well known equation for the dipole-dipole interaction.

Quantum mechanical treatment of the qubit-TLS coupling

Martinis *et al.* published a relation between the qubit-TLS coupling strength υ and the dipole moment component of the TLS $\mu_{\parallel}$ parallel to the $\vec{E}$-field [Mar+05],

$$\upsilon = 2\frac{\mu_{\parallel}}{x}\sqrt{\frac{E_{10}}{2C}}, \tag{5.26}$$

with the junction thickness x, the qubit capacitance C and the qubit transition energy E_{10}. We can understand this equation by the following considerations.

The classical interaction Hamiltonian $H_{\text{q-T}}$ between an electric dipole in a homogeneous electric field $\vec{E}$ is

$$H_{\text{q-T}} = \vec{\mu} \cdot \vec{E} = \mu \cdot \frac{U}{x} \cdot \cos\phi, \tag{5.27}$$

with ϕ being the angle between the dipole and the field direction and U is the voltage at the capacitor with the distance x between the plates. The classical voltage U turns in quantum physics into the operator $\hat{U}$,

$$U = \langle \hat{U} \rangle = \frac{\langle \hat{Q} \rangle}{C} \tag{5.28}$$

with the expectation value of the charge on the capacitor $Q = \langle \hat{Q} \rangle$. For the two lowest states, a phase qubit can be modeled as an harmonic LC-resonator with the Hamiltonian

$$\hat{H}_{\mathrm{LC}} = \frac{\hat{Q}^2}{2C} + \frac{\hat{\Phi}^2}{2L} = E_{10}\left(\hat{a}^\dagger \hat{a} + \frac{1}{2}\right), \tag{5.29}$$

and so

$$\langle \hat{Q} \rangle = i\sqrt{\frac{CE_{10}}{2}}\langle(\hat{a}^\dagger - \hat{a})\rangle. \tag{5.30}$$

For the eigenstates, $\langle n|\hat{Q}|n\rangle = 0$ and therefore there is no (oscillating) electric field in the capacitor. However, a superposition of eigenstates, e.g. $\Big(|n\rangle + i\,|n+1\rangle\Big)/\sqrt{2}$ yields

$$\frac{1}{2}\Big(\langle n| - i\langle n+1|\Big)\left(\hat{a}^\dagger - \hat{a}\right)\Big(|n\rangle + i|n+1\rangle\Big) = -i\sqrt{n+1}, \tag{5.31}$$

and with $n = 0$

$$H_{\mathrm{q-T}} = \frac{\mu}{x}\sqrt{\frac{E_{10}}{2C}}\cos\phi \tag{5.32}$$

For Martinis' Eq. (5.26) a factor 2 is still missing. However, this is only a question of definition. Classically, $H_{\mathrm{q-T}}$ would be the value by which the energy of the qubit-TLS system is decreased or increased due to interaction. However, the coupling strengths used in this thesis, and also by Martinis *et al.*, base on the energy differences in the experiment, which are then $2H_{\mathrm{q-T}} = \upsilon$. For example, $\upsilon_\perp$ denotes the transversal qubit-TLS coupling strength corresponding to the size of the anti-crossing. This means that due to the level repulsion the unperturbed energy was increased or decreased by $\upsilon_\perp/2$. Therefore,

$$\upsilon = 2\frac{\mu\cos\phi}{x}\sqrt{\frac{E_{10}}{2C}} = 2\frac{\mu_\|}{x}\sqrt{\frac{E_{10}}{2C}}, \tag{5.33}$$

where $\mu_{i,\|} = \mu_i\cos\phi$ is the geometrical dipole moment component parallel to the electric fields inside the JJ. υ represents the bare qubit-TLS coupling strength depending on the dipole size and orientation, which have nothing to do with the quantum angles α_i (and β_i) responsible for the TLS asymmetry.

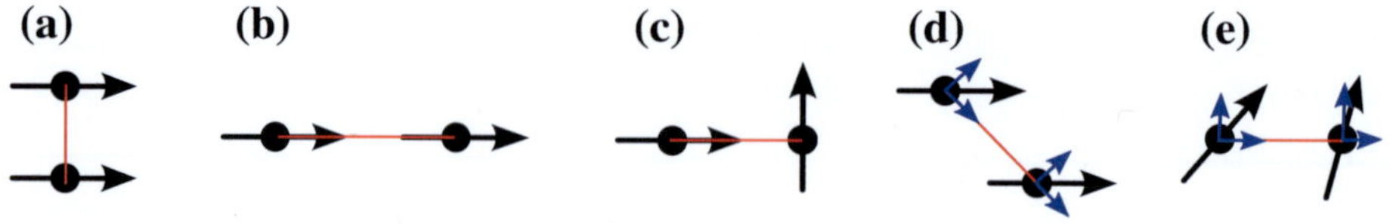

Figure 5.21: Some special configurations of two dipoles are presented. The red connecting lines are parallel to $\vec{r}$. **(a)** and **(b)** show the two main cases of parallel oriented dipoles located perpendicular or parallel to $\vec{r}$, respectively. Although being parallel, (a) corresponds to a repulsive but (b) to an attractive case. The sketches **(c)-(e)** visualize situation with zero interaction energy. As blue arrows show, in (d) and (e) the attractive and repulsive forces annihilate each other if the magic angle condition is fulfilled.

Electric dipole-dipole interaction

Assuming that TLS1 and TLS2 interact directly via their electric dipoles, the relation between the interaction energy and the spatial configuration of the dipoles is known to be

$$H_{\mathrm{dd}} = \frac{1}{4\pi\varepsilon_0\varepsilon_r r^3}\left(\vec{\mu}_1 \cdot \vec{\mu}_2 - \frac{3(\vec{\mu}_1 \cdot \vec{r})(\vec{\mu}_2 \cdot \vec{r})}{\vec{r}^2}\right), \tag{5.34}$$

where $\vec{r}$ is the relative position vector from TLS1 to TLS2. Kocbach *et al.* derived a much nicer representation of this formula [KL10] by decomposing the dipole moments into their parallel and perpendicular components relative to $\vec{r}$,

$$\vec{\mu}_i = \vec{\mu}_{i,\perp} + \vec{\mu}_{i,\parallel}. \tag{5.35}$$

The dipole-dipole interaction can then be rewritten as

$$H_{\mathrm{dd}} = \frac{1}{4\pi\varepsilon_0\varepsilon_r r^3}\left(\vec{\mu}_{1\perp}\vec{\mu}_{2\perp} - 2\vec{\mu}_{1\parallel}\vec{\mu}_{2\parallel}\right). \tag{5.36}$$

Since the TLSs couple to their environment via the charge (or the asymmetry), this potential energy equals $g/2$.

Fig. 5.21 displays some special and interesting cases. In (a) and (b) the dipoles are parallel, but while in (a) the interaction energy is positive, it is negative in (b). (c) shows a trivial case with no interaction. However, there are also non-trivial non-interacting configurations. Assuming both dipoles to be parallel, and going from the repulsive situation in (a) to the attractive in (b), it is obvious that at some angle θ between the dipole direction $\vec{\mu}$ and the connecting vector $\vec{r}$ the interaction strength is zero. Eq. (5.36) yields an obvious equation for this so called magic angle θ_{mag},

$$H_{\text{dd}} \propto \mu_1\mu_2(\sin^2\theta_{\text{mag}} - 2\cos^2\theta_{\text{mag}}) = 0 \tag{5.37}$$

$$\Rightarrow \tan\theta_{\text{mag}} = \sqrt{2} \Leftrightarrow \theta_{\text{mag}} = 54.74°. \tag{5.38}$$

Also if $\vec{\mu}_1 \nparallel \vec{\mu}_2$, a configuration such as in (e), the interaction energy might be zero. For $\theta_1 = 45°$, for example, $\mu_{1,\perp} = \mu_{1,\parallel}$. θ_2 has then to fulfill the relation $\mu_{2,\perp} = 2\mu_{2,\parallel}$ yielding $\theta_2 = 63.43°$.

Some special assumptions on TLS1 and TLS2 dipoles

The main conclusion from the previously discussed paragraph is that there is no minimal distance between the two TLSs. In principle, one could think the closer the dipoles are, the stronger is the interaction. However, if, e.g., the magic angle condition is fulfilled, they can be very close together without feeling each other. Also flipping (or exciting) one TLS does not change this.

To get a very rough estimate of the maximal TLS1-TLS2 distance, there is no other way than making several arbitrary assumptions. The first step is to determine the TLS dipole moment components $\mu_{i,\parallel}$ parallel to the $\vec{E}$-field by using Eq. (5.33). The parameters of the qubit are $x = 2\,\text{nm}$, $C = 850\,\text{fF}$, $E_{01} = 6.3\,\text{GHz}$, yielding

$$\mu_{1,\parallel} = 0.5\ e\text{Å},\ \mu_{2,\parallel} = 0.4\ e\text{Å}. \tag{5.39}$$

If $\mu_i = \mu_{i,\parallel}$, the upper limit for the distances can be calculated by assuming a vertical or horizontal arrangement of the TLSs (Eq. (5.36) and $\varepsilon_r = 10$ (sapphire)). This leads to distances of 6.8 nm or 5.4 nm, respectively. Due to the JJ thickness of 2 nm the first case can be ruled out.

Another meaningful assumption is to take for both dipoles the moments estimated in other publications. Martinis *et al.*, for example, based on Eq. 5.33 and their maximal observed qubit-TLS coupling strength [Mar+05] evaluated a dipole moment of $\mu = 1.3\ e\text{Å}$. A value of about $1e\text{Å}$ was also confirmed in [Col+10] and also theoretically in [DuB+13]; [Aga+13]. Therefore, we take $1e\text{Å}$ as the TLS dipole moments. By using the parallel components of the dipole moment of TLS1 and TLS2 (Eq. (5.39)), we get as the angles between the moments and the electric field

$$\phi_1 = 60° \text{ and } \phi_2 = 66°. \tag{5.40}$$

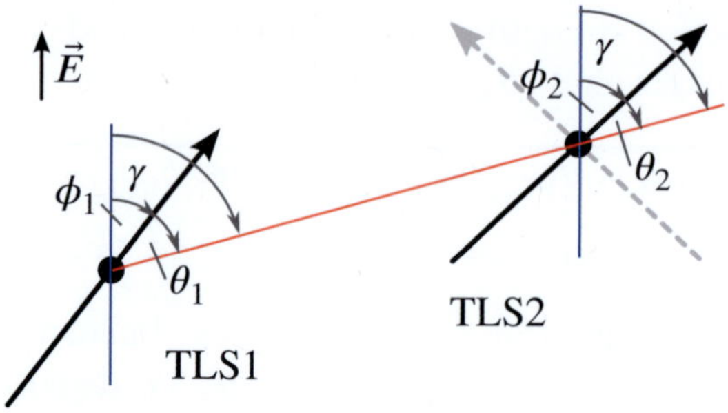

Figure 5.22: The geometry of TLS1 and TLS2 with respect to the electric fields is explained. Although the absolute values of the angles $|\theta_i'|$ are fixed, it is possible to use $\pm\theta_i'$ resulting in $\theta_i = \gamma \pm \theta_i'$. However, there are only two different cases, for example, a positive θ_1' and $\pm\theta_2'$. The maximal distance between TLS1 and TLS2 results to be between 7 and 12 nm.

One could argue that due to the low frequency dependence of TLS1 on strain, TLS1 should be oriented rather perpendicular to the E-field (parallel to the strain change). For TLS2 it would be rather the other way round. However, although the mean strain change is indeed perpendicular to the E-field, locally due to leverage effects and electrical interaction the atoms might be shifted in arbitrary directions, thus completely destroying this argument. In contrast, since the qubit-TLS coupling has to be a purely electric interaction, the E-field represents a distinct direction for the moments.

To maximize the interaction, we assume the TLS1 and TLS2 dipoles to lie in one plane (Fig. 5.22). We define the angle γ as the angle between $\vec{r}$ and $\vec{E}$, so that the main angle for the dipole-dipole interaction, θ_i, is simply replaced by $\theta_i = \gamma \pm \phi_i$. Without loss of generality, we set $\theta_1 = \gamma - \phi_1$ and consider only the two cases $\theta_2 = \gamma \pm \phi_2$, as shown in Fig. 5.22. The four trivial configurations with $\gamma = 0$ and $\gamma = 90°$ yield as the distance between TLS1 and TLS2 values between 7 nm and 11 nm, so that again the arrangement where $\vec{r} \parallel \vec{E}$ is not realistic. The absolutely maximal distance occurs when TLS1 and TLS2 show approximately in the same direction, $\gamma = 63°$, resulting in a distance of 12 nm. However, this case is not really possible, because the displacement of the two TLSs in direction of the electric field exceeds the JJ thickness.

Altogether, one can conclude that the distance of the two TLSs is in the range of about $\approx$ 10 nm. Distributing all "visible" $\approx$ 50 TLSs in the qubit spectrum on our 1 μm^2 large junction, the average area for one TLS is about 140×140 nm^2. Therefore, the estimated maximal distance makes insofar sense as it really corresponds to a quite rare case to observe two coherently coupled TLSs with a phase qubit.

6 Conclusion

The aim of my PhD thesis was to link two completely different fields of physics. Our group, including me and, in particular, my supervisor during my diploma thesis Jürgen Lisenfeld, were working with superconducting quantum bits (qubits), while the PhD student Torben Peichl from the group of Prof. Georg Weiss investigated the two-level tunneling systems (TLSs) in solids. We started this project together. Torben Peichl finished his PhD approximately one year ago and Jürgen and I continued to take data and improve the experiment.

Qubits are controllable quantum devices, which may become very helpful in solving complicated mathematical problems and better understanding the physics of the microcosm. However, a lot of fundamental research has to be still done. As it is always the case in quantum mechanics, decoherence effects are the main problem. Jointly with low temperatures, clever engineering and low-noise electronics, the used materials play also a major role. In superconducting qubits, the insulating layer inside the Josephson junction (JJ), usually AlO_x, is blamed to contribute to the qubit decoherence rate significantly, mainly due to dielectric losses. Furthermore, it is assumed that inside of this disordered tunneling barrier parasitic quantum systems are located, which couple coherently to the qubit and induce avoided level crossings in its spectrum. In particular, the spectra of phase qubits show many anti-crossings due to their relatively large JJ.

The two-level tunneling systems (TLSs) constitute the basis of the tunneling model. They explain the large discrepancies between the characteristics of amorphous solids and crystals at low temperatures. TLSs are defects where atoms or small groups of atoms tunnel between two stable configurations, which is described by a double well potential. A key feature of this model, which is also one of the most important assumptions in this thesis, is that the elastic deformation of the crystal changes only the asymmetry energy ε, which is defined as the energy difference between the two wells. The tunneling energy Δ, which depends on other parameters like the barrier height or the tunneling distance, remains constant. The excitation energy has the form $E = \sqrt{\varepsilon^2 + \Delta^2}$

and therefore has a hyperbolic dependence on strain. Far outside the ''sweet point'', $|\varepsilon| \gg |\Delta|$, the hyperbola turns into a linear function of strain.

Our aim was to verify experimentally, whether the parasitic quantum systems seen by the superconducting qubit behave the way predicted for TLSs when applying strain fields. Therefore, we constructed a sample holder, in which a built-in piezo actuator generates force on the chip and thus slightly deforms it. Our estimations and simulations yield a deformation value of $8 \cdot 10^{-7}$ per 1 V on the piezo, which corresponds to a relative length change of an atomic nucleus per 1 nm. We studied the TLSs in two qubits from different generations. The qubit from the first generation of flux-biased phase qubits has rather poor coherence times, but therefore shows many anti-crossings in its spectrum. The second sample has much fewer anti-crossings, however, its coherence times allow one to take time domain measurements, which significantly improve the results and reduce the measuring time. Via a special protocol, the resonance frequencies of the parasitic systems could be tracked as a function of the applied strain. The resulting spectroscopy images of both samples are qualitatively similar. Both qubits reveal that the resonance frequencies of almost all parasitic systems depend linearly or hyperbolically on strain, as predicted for TLSs by the tunneling model. Sometimes, abrupt frequency jumps can be observed. With regard to those TLS traces that show a jump within the frequency range accessible by the phase qubit, we never detected a change of the resonance frequency dependence on strain other than a shift. We could even observe ''double traces'' between which the parasitic system is constantly switching back and forth. These phenomena are explained in terms of the tunneling model in the following way: In the vicinity of the coherent TLS coupled to the qubit another, incoherent TLS might be located which modifies the resonance frequency of the first one and which switches only once per hour, day, month etc.

If the second TLS does not flip abruptly, but one is able to control the ratio of the positional probabilities in the two wells by varying the strain field, one expects the first TLS to show a smooth change of its resonance frequency. We could study exactly such an S-shape in detail. Furthermore, the trace of TLS1, which is strongly coupled to the qubit, shows an anti-crossing on each side of the S-shape. This proves that TLS1 is coherently interacting with TLS2, which is, however, invisible on the qubit spectrum. On the one hand, the observed structures can be explained classically by assuming two interacting electric dipoles. Smoothly flipping the second dipole shifts the energy of the first one resulting in the S-shape. If both dipoles are in resonance, rotations of one dipole are transferred to the other and vice versa, which induces the anti-crossings. On the other hand, one can use the predictions of the tunneling model. As it is shown in this thesis, the data are qualitatively as well as quantitatively reproduced with very high accuracy. However, due to the large number of parameters, we performed an additional experiment. With the help of my simulation accounting for three quantum systems, we succeeded to optimize the pulse sequence and to measure the energy of the full excited TLS1-TLS2 system at the TLS1-TLS2 resonance. This allowed us to determine several parameters directly and thus to improve the fitting procedure significantly. Further investigation of the TLS1-TLS2 anti-crossing indicated a different visibility of the two branches. This could be explained by assuming an existing but very small direct qubit-TLS2 coupling, which decreases the interaction between the qubit and the upper branch but enhances the coupling between the qubit and the lower branch. One of the main results is that TLS2 is not invisible to the qubit due to a particular orientation of its dipole moment or its value. On the contrary, the bare qubit-TLS2 coupling is approximately as strong as that between the qubit and TLS1. However, the tunneling energy of TLS2 is so small, that at qubit frequencies its potential is very asymmetric, which dramatically decreases the observable qubit-TLS2 coupling strength. Finally, we made a very rough estimation of the TLS1-TLS2 distance. We assumed the TLS1-TLS2 interaction to be purely electric and analyzed for some particular TLS1 and TLS2 orientations the formula describing the dipole-dipole interaction. On the one hand, there is no minimal distance. This means, that for a particular situation two dipoles can be located arbitrarily close together without feeling each other. On the other hand, the maximum distance was estimated to be of the order of 10 nm. If we distribute all visible ≈ 50 TLSs on the 1 μm^2 large Josephson junction, one TLS can be found in an area of 140x140 nm^2. These numbers confirm that it is a very rare case to observe to coherently interacting TLSs with a phase qubit on average.

7 Zusammenfassung

Das Ziel meiner Doktorarbeit war es, zwei auf den ersten Blick komplett unterschiedliche physikalische Bereiche zu verknüpfen. Auf der einen Seite werden schon seit über 40 Jahren Defekte in Gläsern studiert. Manche Fehlstellen können instabil sein, so dass einzelne Atome zwischen zwei Positionen tunneln und somit Zwei-Zustands-Quantensysteme bilden können, so genannte TLS (two-level tunneling system). Auf der anderen Seite entwickelt man bereits seit ca. zehn Jahren die Technologie, einzelne und kohärente Zwei-Zustands-Quantensysteme auf Basis supraleitender Schaltungen zu konstruieren und zu kontrollieren (superconducting qubits). Interessanterweise beobachtet man bei diesen Systemen eine kohärente Wechselwirkung mit anderen Quantensystemen, deren mikroskopischer Ursprung noch nicht vollständig geklärt ist. Durch ein ausgeklügeltes Experiment ist es uns gelungen zu zeigen, dass sich diese mysteriösen Quantensysteme im Bezug auf mechanische Spannung genauso verhalten wie es von tunnelnden Atomen erwartet wird.

Die Quantenbits, oder Qubits, sind die Analogons der klassischen Bits in der Quantenwelt. Es gibt Theorien, die belegen, dass ein Quantencomputer tatsächlich einige wichtige mathematische Probleme in viel kürzerer Zeit lösen kann als ein herkömmlicher Computer. Abgesehen von dieser Zukunftsvision sind Qubits ausgezeichnete Objekte, um die Quantenmechanik tiefer zu ergründen und um die Dekohärenz, die die eigentliche Verknüpfung zwischen der Quanten- und der klassischen Physik darstellt, besser zu verstehen.

Eines der erfolgversprechendsten Konzepte für die Basis eines Quantencomputers sind elektronische Schaltkreise aus supraleitenden Materialien. Dieser Ansatz ermöglicht sowohl eine sehr einfache Kontrolle des Qubits als auch die Kopplung von mehreren Qubits. Allerdings sind sehr tiefe Temperaturen erforderlich, zu deren Erzeugung Mischungskryostaten eingesetzt werden, die den Chip auf bis zu 30 mK abkühlen. Die Funktionsweise von supraleitenden Qubits basiert auf dem Effekt des Josephson-Kontaktes (JK). Dies ist eine sehr dünne Isolatorschicht (~ 2 nm), die den Supraleiter unterbricht. Der maximale Suprastrom durch einen JK ist zwar klein, der JK stellt aber aufgrund seiner nicht-linearen Strom-Spannungs-Kennlinie ein komplett neuartiges elektronisches Bauteil dar. Es gibt verschiedene Typen von supraleitenden Qubits, die den JK in jeweils unterschiedlichen Regimes verwenden. Die Experimente in dieser Doktorarbeit wurden mit einem Phasenqubit durchgeführt. Ein Phasenqubit ist eine supraleitende Schleife, die durch einen JK unterbrochen ist. Mit Mikrowellenpulsen, die auf das Qubit geschickt werden, kann die Eigenfrequenz gefunden und der Qubitzustand manipuliert werden und mittels eines angelegten externen Flusses wird die Resonanzfrequenz des Phasen-Qubits kontrolliert. Die Spektren von Phasenqubits als Funktion des externen Flusses weisen Unterbrechungen bzw. Aufspaltungen bei bestimmten Frequenzen auf. Die Erklärung hierfür sind mikroskopische Zweiniveausysteme, die kohärent mit dem Qubit wechselwirken. Die eigentliche Natur dieser Systeme ist noch nicht vollständig erforscht. Allerdings gibt es eindeutige Hinweise, dass sie innerhalb der Tunnelbarriere des JK lokalisiert sind. Deshalb ist es das Nächstliegendste anzunehmen, das es sich hierbei um die TLS handelt. Das Hauptargument dieser Annahme ist, dass die Isolatorschicht eine ähnlich ungeordnete Struktur hat wie es bei den Gläsern der Fall ist.

In amorphen Materialien wurden die TLS erst Anfang der 70er Jahre als solche identifiziert. Davor galt das Debyemodell als äußerst erfolgreich in der Beschreibung der Wärmeeigenschaften von Festkörpern. Allerdings fanden 1971 Zeller und Pohl heraus, dass die Eigenschaften von amorphen Festkörpern bei Temperaturen unter 1 K sehr stark von dem Debyemodell abweichen. Ein Jahr später entwickelten Phillips und unabhängig auch Anderson *et al.* das Tunnelmodell. Demnach gibt es einzelne Defekte, bei denen manche Atome bzw. Atomgruppen zwischen zwei stabilen Konfigurationen, die über eine kleine Übergangsrate miteinander gekoppelt sind, tunneln können. Ausgehend davon ist es nicht mehr verwunderlich, dass bei Gläsern, aufgrund einzelner Anregungen der TLS, die spezifische Wärme und die Phononenstreurate um Ordnungen größer ist. Bei geladenen TLS unterscheiden sich auch die dielektrischen Eigenschaften erheblich von denen der kristallinen Festkörper. In den letzten 40 Jahren der Forschung wurde das Tunnelmodel nicht nur durch unzählige Messungen untermauert, sondern auch ein universelles Verhalten amorpher Materialien bei tiefen Temperaturen nachgewiesen.

Auch wenn damals nur Erwartungswerte von einer sehr großen Anzahl und auch unterschiedlicher TLS zugänglich waren, basiert das Tunnelmodel auf einem einzelnen Tunnelsystem. Dieses wird durch ein Doppelmuldenpotetial beschrieben, bei dem die zwei wichtigsten Größen die Asymmetrieenergie ε und die Tunnelenergie Δ sind. Die Energiedifferenz der Eigenzustände beträgt

$E = \sqrt{\Delta^2 + \varepsilon^2}$. Das Tunnelmodel liefert zwei Kopplungsmechanismen zwischen einem TLS und der Umgebung: über elektrische Felder, z.B. mit Hilfe eines äußeren Kondensators, und über elastische Felder, welche z.B. durch die Phononen oder auch durch eine externe Deformation der Probe erzeugt werden. Dem Tunnelmodel zufolge verändern elektrische und elastische Felder nur die Asymmetrieenergie, die Tunnelenergie bleibt erhalten. Da weiterhin angenommen wird, dass sich ε linear mit der Feldstärke ändert, ergibt sich eine hyperbolische Abhängigkeit der Anregungsenergie E als Funktion von der Feldstärke, die im Falle von $|\varepsilon| \gg |\Delta|$ in eine lineare Funktion mündet.

Durch das Spektroskopieren des Phasenqubits lassen sich die Resonanzen der TLS vermessen. Wie bereits erwähnt, erwarten wir eine Wechselwirkung der TLS mit äußeren elektrischen oder elastischen Feldern. In unserem Fall ist die Wechselwirkung zwischen dem TLS und dem elektrischen Feld verantwortlich für die Qubit-TLS-Kopplung, denn der JK stellt aufgrund seiner Geometrie einen Kondensator dar. Eine Erzeugung externer elektrischer Felder kommt nicht in Frage, da der JK des Phasenqubits über einen Supraleiter kurzgeschlossen ist und deshalb äußere elektrische Felder abgeschirmt werden. Allerdings lässt sich eine mechanische Deformation des JK bewerkstelligen. Mittels eines speziell angefertigten Probenträgers mit einem eingebauten Piezostellenelement haben wir den Chip verbogen. Unsere Abschätzungen liefern eine Deformation von $8 \cdot 10^{-7}$ pro anliegendes Volt am Piezostellenelement, was einer Verlängerung eines Nanometers um einen Atomkerndurchmesser entspricht. Wie erwartet zeigen die TLS eine überwiegend lineare oder hyperbolische Resonanzabhängigkeit von der Deformation. Ab und zu können wir abrupte Änderungen der TLS-Frequenzen beobachten. Allerdings stellen wir fest, dass, sofern sich der Sprung innerhalb des mit dem Qubit messbaren Frequenzfensters ereignet, die Deformationsabhängigkeit (bis auf eine Verschiebung) erhalten bleibt. Falls solch ein TLS sehr schnell zwischen zwei Frequenzen hin und her springt, ergibt sich eine Doppelspur, welche wir auch beobachten konnten. Dieses Phänomen lässt sich wunderbar mit dem Tunnelmodel erklären. In der direkten Nachbarschaft des kohärenten TLS (TLS1), das an das Qubit koppelt, kann es ein weiteres, inkohärentes TLS geben (TLS2), das seinen Zustand nur ab und zu ändert und somit einen Sprung der Resonanzfrequenz von TLS1 bewirkt.

Eine sehr interessante Situation ergibt sich, wenn man so ein gekoppeltes TLS1-TLS2-System hat, bei dem allerdings auch TLS2 sehr stark deformationsabhängig ist, so dass man den Schwerpunkt seiner Grundzustandswellenfunktion kontrolliert von einer Mulde in die andere verschieben kann. Man rechnet dann nicht mit einem plötzlichen Sprung sondern mit einer glatten Verschiebung der TLS1-Resonanz. In der Tat haben wir so eine „S“-förmige Spur beobachtet, die sich sowohl qualitativ als auch quantitativ mit dem TLS1-TLS2-Modell erklären lässt. Des Weiteren weist die Spur symmetrisch um das „S“ jeweils eine Niveauaufspaltung auf. Diese ist auf eine kohärente Wechselwirkung zwischen den beiden TLS zurückzuführen. Eine vollständige Analyse des Systems wird dadurch erschwert, dass TLS1 nahezu unabhängig von der Deformation und dass die Spur von TLS2 auf dem Spektrum kaum zu erkennen ist. Da die quantenmechanische Beschreibung viele Parameter beinhaltet, war es unumgänglich durch ein zusätzliches und kompliziertes Experiment die Energie des vollständig angeregten TLS1-TLS2-Systems zu messen. Dies ermöglichte es, einige Größen direkt zu extrahieren und somit die Fittingprozedur entscheidend zu verbessern. Des Weiteren konnten wir noch eine sehr kleine direkte Qubit-TLS2-Kopplung nachweisen. Es stellt sich heraus, dass die „reine“ Qubit-TLS2-Kopplung fast genauso groß ist wie die zwischen Qubit und TLS1. Allerdings verringert sich die Kopplungsstärke erheblich, wenn das TLS Potential stark unsymmetrisch ist. Dies ist bei TLS2 im Bereich der Qubitfrequenzen der Fall. Abschließend haben wir versucht, den Abstand zwischen TLS1 und TLS2 zu bestimmen, indem wir von einer reinen elektrischen Wechselwirkung und der Dipol-Dipol-Formel ausgingen. Auf der einen Seite ergibt sich, dass es keinen Mindestabstand gibt. Das bedeutet, dass zwei Dipole bei einer bestimmten Ausrichtung beliebig nahe lokalisiert sein können, ohne sich gegenseitig zu beeinflussen. Auf der anderen Seite liegt die Maximaldistanz in der Größenordnung von 10 nm. Wenn wir alle sichtbaren $\approx$ 50 TLS gleichmäßig im 1 μm^2 großen JK verteilen, dann findet man jeweils ein TLS in einem Bereich der Größe 140x140 nm^2. Diese Abschätzung bestätigt, dass es ein sehr seltener Fall ist, zwei kohärente und gekoppelte TLS zu beobachten.

Bibliography

[Aga+13] Kartiek Agarwal, Ivar Martin, Mikhail D. Lukin, and Eugene Demler. "Polaronic model of two-level systems in amorphous solids". In: *Phys. Rev. B* 87 (2013), p. 144201.

[AHV72] P. W. Anderson, B. I. Halperin, and C. M. Varma. "Anomalous low-temperature thermal properties of glasses and spin glasses". In: *Philosophical Magazine* 25 (1972), p. 1.

[AH75] W. Arnold and S. Hunklinger. "Experimental evidence for the direct interaction between two-level systems in glasses at very low temperatures". In: *Solid State Communications* 17 (1975), p. 883.

[BP82] Antonio Barone and Gianfranco Paterno. *Physics and Applications of the Josephson Effect*. Wiley-VCH, 1982.

[BH77] J. L. Black and B. I. Halperin. "Spectral diffusion, phonon echoes, and saturation recovery in glasses at low temperatures". In: *Phys. Rev. B* 16 (1977), p. 2879.

[Bus+11] P. Bushev, A. K. Feofanov, H. Rotzinger, I. Protopopov, J. H. Cole, C. M. Wilson, G. Fischer, A. Lukashenko, and A. V. Ustinov. "Ultralow-power spectroscopy of a rare-earth spin ensemble using a superconducting resonator". In: *Phys. Rev. B* 84 (2011), p. 060501.

[Bus+10] P. Bushev, C. Müller, J. Lisenfeld, J. H. Cole, A. Lukashenko, A. Shnirman, and A. V. Ustinov. "Multiphoton spectroscopy of a hybrid quantum system". In: *Phys. Rev. B* 82 (2010), p. 134530.

[Chi+03] I. Chiorescu, Y. Nakamura, C. J. P. M. Harmans, and J. E. Mooij. "Coherent Quantum Dynamics of a Superconducting Flux Qubit". In: *Science* 299 (2003), p. 1869.

[CB04] John Clarke and Alex I. Braginski. *The SQUID Handbook: Fundamentals and Technology of SQUIDs and SQUID Systems*. Wiley-VCH; 1 edition, 2004.

[Cla+94] J. Classen, C. Enss, C. Bechinger, G. Weiss, and S. Hunklinger. "Low frequency acoustic and dielectric measurements on glasses". In: *Annalen der Physik* 506 (1994), p. 315.

[CDL06] Claude Cohen-Tannoudji, Bernard Diu, and Frank Laloe. *Quantum Mechanics, Volume 1.* Wiley-Interscience, 2006.

[CAG09] J. H. Cole, J. C. Ang, and A. D. Greentree. "Solving super operator problems". Unpublished. Centre of Quantum Computer Technology, School of Physics, University of Melbourne, Victoria 3010, Australia. 2009.

[Col+10] J. H. Cole, C. Müller, P. Bushev, G. J. Grabovskij, J. Lisenfeld, A. Lukashenko, A. V. Ustinov, and A. Shnirman. "Quantitative evaluation of defect-models in superconducting phase qubits". In: *Applied Physics Letters* 97 (2010), p. 252501.

[DWM04] M. H. Devoret, A. Wallraff, and J. M. Martinis. "Superconducting Qubits: A Short Review". In: *ArXiv Condensed Matter e-prints* (2004). eprint: `arXiv:cond-mat/0411174`.

[DCL06] Bernard Diu, Claude Cohen-Tannoudji, and Frank Laloe. *Quantum Mechanics, Volume 2.* Wiley-Interscience, 2006.

[DiV00] David P. DiVincenzo. "The Physical Implementation of Quantum Computation". In: *Fortschritte der Physik* 48 (2000), p. 771.

[DuB+13] Timothy C. DuBois, Manolo C. Per, Salvy P. Russo, and Jared H. Cole. "Delocalized Oxygen as the Origin of Two-Level Defects in Josephson Junctions". In: *Physical Review Letters* 110 (2013), p. 077002.

[EH00] C. Enss and S. Hunklinger. *Tieftemperaturphysik*. Springer Singapore Pte. Limited, 2000.

[FI07] Lara Faoro and Lev B. Ioffe. "Microscopic origin of critical current fluctuations in large, small, and ultra-small area Josephson junctions". In: *Phys. Rev. B* 75 (2007), p. 132505.

[Gra+11] G. J. Grabovskij, P. Bushev, J. H. Cole, C. Müller, J. Lisenfeld, A. Lukashenko, and A. V. Ustinov. "Entangling microscopic defects via a macroscopic quantum shuttle". In: *New Journal of Physics* 13 (2011), p. 063015.

[Gra10] Grigorij Grabovskij. “Coherent manipulation of microscopic defects in a superconducting phase qubit”. MA thesis. Friedrich-Alexander-Universität Erlangen-Nürnberg, 2010.

[Gra+12] Grigorij J. Grabovskij, Torben Peichl, Jürgen Lisenfeld, Georg Weiss, and Alexey V. Ustinov. “Strain Tuning of Individual Atomic Tunneling Systems Detected by a Superconducting Qubit”. In: *Science* 338 (2012), p. 232.

[Hah50] E. L. Hahn. “Spin Echoes”. In: *Phys. Rev.* 80 (1950), p. 580.

[HPP68] J. P. Harrison, P. P. Peressini, and R. O. Pohl. “Study of an Impurity Mode Using Specific-Heat Measurements”. In: *Phys. Rev.* 171 (1968), p. 1037.

[Hun07] S. Hunklinger. *Festkörperphysik*. Oldenbourg, 2007.

[Ins] Oxford Instruments. url: `http://www.oxford-instruments.com`.

[JC63] E. T. Jaynes and F. W. Cummings. “Comparison of quantum and semiclassical radiation theories with application to the beam maser”. In: *Proceedings of the IEEE* 51 (1963), p. 89.

[Joh+05] Philip R. Johnson, William T. Parsons, Frederick W. Strauch, J. R. Anderson, Alex J. Dragt, C. J. Lobb, and F. C. Wellstood. “Macroscopic Tunnel Splittings in Superconducting Phase Qubits”. In: *Phys. Rev. Lett.* 94 (2005), p. 187004.

[Jos74] B. D. Josephson. “The discovery of tunnelling supercurrents”. In: *Rev. Mod. Phys.* 46 (1974), p. 251.

[Kli+09] Jeffrey S. Kline, Haohua Wang, Seongshik Oh, John M. Martinis, and David P. Pappas. “Josephson phase qubit circuit for the evaluation of advanced tunnel barrier materials”. In: *Superconductor Science and Technology* 22 (2009), p. 015004.

[KL10] Ladislav Kocbach and Suhail Lubbad. “Geometrical simplification of the dipole-dipole interaction formula”. In: *Physics Education* 45 (2010), p. 345.

[Kru+06] J. Krupka, J. Breeze, A. Centeno, N. Alford, T. Claussen, and L. Jensen. “Measurements of Permittivity, Dielectric Loss Tangent, and Resistivity of Float-Zone Silicon at Microwave Frequencies”. In: *Microwave Theory and Techniques, IEEE Transactions on* 54 (2006), p. 3995.

[KY05] Li-Chung Ku and Clare C. Yu. “Decoherence of a Josephson qubit due to coupling to two-level systems”. In: *Phys. Rev. B* 72 (2005), p. 024526.

[Kul11] H. J. Kull. *Laserphysik: Physikalische Grundlagen des Laserlichts und seine Wechselwirkung mit Materie*. Oldenbourg Wissenschaftsverlag, 2011.

[LL81] L. D. Landau and L. M. Lifshitz. *Quantum Mechanics (Non-Relativistic Theory) Volume 3*. Butterworth-Heinemann (3 edition), 1981.

[Las+75] J. C. Lasjaunias, A. Ravex, M. Vandorpe, and S. Hunklinger. “The density of low energy states in vitreous silica: Specific heat and thermal conductivity down to 25 mK”. In: *Solid State Communications* 17 (1975), p. 1045.

[Lin76] G. Lindblad. “On the generators of quantum dynamical semigroups”. In: *Communications in Mathematical Physics* 48 (1976), p. 119.

[Lis08] Jürgen Lisenfeld. “Experiments on Superconducting Josephson Phase Quantum Bits”. PhD thesis. Friedrich-Alexander-Universität Erlangen-Nürnberg, 2008.

[Lis+10a] Jürgen Lisenfeld, Clemens Müller, Jared H. Cole, Pavel Bushev, Alexander Lukashenko, Alexander Shnirman, and Alexey V. Ustinov. “Rabi spectroscopy of a qubit-fluctuator system”. In: *Phys. Rev. B* 81 (2010), p. 100511.

[Lis+10b] Jürgen Lisenfeld, Clemens Müller, Jared H. Cole, Pavel Bushev, Alexander Lukashenko, Alexander Shnirman, and Alexey V. Ustinov. “Rabi spectroscopy of a qubit-fluctuator system”. In: *Phys. Rev. B* 81 (2010), p. 100511.

[Mar09] John Martinis. “Superconducting phase qubits”. In: *Quantum Information Processing* 8 (2009), p. 81.

[Mar+05] John M. Martinis, K. B. Cooper, R. McDermott, Matthias Steffen, Markus Ansmann, K. D. Osborn, K. Cicak, Seongshik Oh, D. P. Pappas, R. W. Simmonds, and Clare C. Yu. “Decoherence in Josephson Qubits from Dielectric Loss”. In: *Phys. Rev. Lett.* 95 (2005), p. 210503.

[MK89] John M. Martinis and R. L. Kautz. "Classical phase diffusion in small hysteretic Josephson junctions". In: *Phys. Rev. Lett.* 63 (1989), p. 1507.

[Mar+02] John M. Martinis, S. Nam, J. Aumentado, and C. Urbina. "Rabi Oscillations in a Large Josephson-Junction Qubit". In: *Phys. Rev. Lett.* 89 (2002), p. 117901.

[McC68] D. E. McCumber. "Effect of ac Impedance on dc Voltage-Current Characteristics of Superconductor Weak-Link Junctions". In: *Journal of Applied Physics* 39 (1968), p. 3113.

[MD01] Rémy Mosseri and Rossen Dandoloff. "Geometry of entangled states, Bloch spheres and Hopf fibrations". In: *Journal of Physics A: Mathematical and General* 34 (2001), p. 10243.

[MSM09] Clemens Müller, Alexander Shnirman, and Yuriy Makhlin. "Relaxation of Josephson qubits due to strong coupling to two-level systems". In: *Phys. Rev. B* 80 (2009), p. 134517.

[NPT99] Y. Nakamura, Yu. A. Pashkin, and J. S. Tsai. "Coherent control of macroscopic quantum states in a single-Cooper-pair box". In: *Nature* 398 (1999), p. 786.

[Nee+08] Matthew Neeley, M. Ansmann, Radoslaw C. Bialczak, M. Hofheinz, N. Katz, Erik Lucero, A. O'Connell, H. Wang, A. N. Cleland, and John M. Martinis. "Process tomography of quantum memory in a Josephson-phase qubit coupled to a two-level state". In: *Nat Phys* 4 (2008), p. 523.

[NC00] Michael A. Nielsen and Isaac L. Chuang. *Quantum Computation and Quantum Information*. Cambridge University Press, 2000.

[Pal+13] E. Paladino, Y. M. Galperin, G. Falci, and B. L. Altshuler. "1/f noise: implications for solid-state quantum information". In: *ArXiv e-prints* (2013). eprint: `1304.7925`.

[Pei12] T. Peichl. "Einfluss mechanischer Deformation auf atomare Tunnelsysteme - untersucht mit Josephson Phasen-Qubits". PhD thesis. Karlsruhe Institute for Technology, 2012.

[Phi72] W.A. Phillips. "Tunneling states in amorphous solids". In: *Journal of Low Temperature Physics* 7 (1972), p. 351.

[Phi81] W.A. Phillips. *Amorphous solids: low-temperature properties*. Springer-Verlag, 1981.

[PI] Physik Instrumente (PI). *Fundamentals of Piezoelectric Actuators.* url: http://www.piezo.ws/piezoelectric_actuator_tutorial/Piezo_Design_part3.php.

[Pre78] William H. Press. In: *Comments on Astrophysics* 7 (1978), p. 103.

[Sch+02] V. V. Schmidt, P. Müller, A. V. Ustinov, and I. V. Grigorieva. *The Physics of Superconductors: Introduction to Fundamentals and Applications.* Springer; 1 edition, 2002.

[SZ97] Marlan O. Scully and M. Suhail Zubairy. *Quantum Optics.* Cambridge University Press, 1997.

[Sen+08] S. Sendelbach, D. Hover, A. Kittel, M. Mück, John M. Martinis, and R. McDermott. "Magnetism in SQUIDs at Millikelvin Temperatures". In: *Phys. Rev. Lett.* 100 (2008), p. 227006.

[Sha+10] Yoni Shalibo, Ya'ara Rofe, David Shwa, Felix Zeides, Matthew Neeley, John M. Martinis, and Nadav Katz. "Lifetime and Coherence of Two-Level Defects in a Josephson Junction". In: *Physical Review Letters* 105 (2010), p. 177001.

[Shn+05] Alexander Shnirman, Gerd Schön, Ivar Martin, and Yuriy Makhlin. "Low- and High-Frequency Noise from Coherent Two-Level Systems". In: *Phys. Rev. Lett.* 94 (2005), p. 127002.

[Sim+04] R. W. Simmonds, K. M. Lang, D. A. Hite, S. Nam, D. P. Pappas, and John M. Martinis. "Decoherence in Josephson Phase Qubits from Junction Resonators". In: *Phys. Rev. Lett.* 93 (2004), p. 077003.

[Ska10] Sebastian T. Skacel. "Investigation of Dielectric Losses in Amorphous Thin Films". MA thesis. Karlsruhe Institute for Technology, 2010.

[Sou+09] Rogério de Sousa, K. Birgitta Whaley, Theresa Hecht, Jan von Delft, and Frank K. Wilhelm. "Microscopic model of critical current noise in Josephson-junction qubits: Subgap resonances and Andreev bound states". In: *Phys. Rev. B* 80 (2009), p. 094515.

[Ste+06] Matthias Steffen, M. Ansmann, R. McDermott, N. Katz, Radoslaw C. Bialczak, Erik Lucero, Matthew Neeley, E. M. Weig, A. N. Cleland, and John M. Martinis. "State Tomography of Capacitively Shunted Phase Qubits with High Fidelity". In: *Phys. Rev. Lett.* 97 (2006), p. 050502.

[Ste68] W. C. Stewart. "Current-voltage characteristics of Josephson junctions". In: *Applied Physics Letters* 12 (1968), p. 277.

[Tay+06] R. P. Taylor, G. F. Nellis, S. A. Klein, D. W. Hoch, J. Fellers, P. Roach, J. M. Park, and Y. Gianchandani. "Measurements of the material properties of a laminated piezoelectric stack at cryogenic temperatures". In: *AIP Conference Proceedings* 824 I (2006), p. 200.

[VH77] M. Von Schickfus and S. Hunklinger. "Saturation of the dielectric absorption of vitreous silica at low temperatures". In: *Physics Letters A* 64 (1977), p. 144.

[Wal01] A. Wallraff. "Fluxon Dynamics in Annular Josephson Junctions: From Relativistic Strings to Quantum Particles". PhD thesis. Friedrich-Alexander-Universität Erlangen-Nürnberg, 2001.

[Web10] Graham A. Webb. *Nuclear Magnetic Resonance*. Royal Society of Chemistry; 1st edition, 2010.

[WL84] K. B. Whaley and J. C. Light. "Rotating-frame transformations: A new approximation for multiphoton absorption and dissociation in laser fields". In: *Phys. Rev. A* 29 (1984), p. 1188.

[ZP71] R. C. Zeller and R. O. Pohl. "Thermal Conductivity and Specific Heat of Noncrystalline Solids". In: *Phys. Rev. B* 4 (1971), p. 2029.

[Zim06] Prof. Dr.Claus Zimmermann. *Quantenoptik*. 2006. url: http://www.pit.physik.uni-tuebingen.de/zimmermann/lehre/skripten/Quantenoptik.pdf.

My publications

Papers

- G. J. Grabovskij, L. J. Swenson, O. Buisson, C. Hoffmann, A. Monfardini, and J. C. Villégier. ''In situ measurement of the permittivity of helium using microwave NbN resonators''. In: *Appl. Phys. Lett.* 93 (2008), p. 134102
- J. H. Cole, C. Müller, P. Bushev, G. J. Grabovskij, J. Lisenfeld, A. Lukashenko, A. V. Ustinov, and A. Shnirman. ''Quantitative evaluation of defect-models in superconducting phase qubits''. In: *Appl. Phys Lett.* 97 (2010), p. 252501
- G. J. Grabovskij, P. Bushev, J. H. Cole, C. Müller, J. Lisenfeld, A. Lukashenko and A. V. Ustinov. ''Entangling microscopic defects via a macroscopic quantum shuttle''. In: *New J. Phys.* 13 (2011), p. 063015
- G. J. Grabovskij, T. Peichl, J. Lisenfeld, G. Weiss, and A. V. Ustinov. ''Strain Tuning of Individual Atomic Tunneling Systems Detected by a Superconducting Qubit''. In: *Science* 338 (2012), p. 232
- G. J. Grabovskij, J. Lisenfeld, C. Müller, J. H. Cole, G. Weiss and A. V. Ustinov. ''Detection of directly interacting coherent two-level systems in a solid''. In preparation.

Conferences

- G. J. Grabovskij , P. Bushev, J. H. Cole, C. Müller, J. Lisenfeld, A. Lukashenko and A. V. Ustinov. ''Experiments with a phase qubit and two-level-fluctuators''. Poster. Deutsche Physikalische Gesellschaft (DPG), Regensburg, Germany, 2010.
- G. J. Grabovskij , P. Bushev, J. H. Cole, C. Müller, J. Lisenfeld, A. Lukashenko and A. V. Ustinov. ''Coherent dynamics of two microscopic defect states coupled via a phase qubit''. Talk. DPG, Dresden, Germany, 2011.

- J. Lisenfeld, G. J. Grabovskij, P. Bushev, A. Lukashenko and A. V. Ustinov, J. H. Cole, C. Müller and A. Shnirman. ''Two-level defects in Josephson qubits''. Poster. International Conference on Quantum Technologies, Moscow, Russia, 2011.
- G. J. Grabovskij, T. Peichl, J. Lisenfeld, G. Weiss, and A. V. Ustinov. ''Coherent atomic tunneling tuned by strain in a Josephson phase qubit''. Talk. SOLID workshop, Delft, Netherlands, 2012.
- G. J. Grabovskij, T. Peichl, J. Lisenfeld, G. Weiss, and A. V. Ustinov. ''Coherent atomic tunneling tuned by strain in a Josephson phase qubit''. Talk. DPG, Berlin, Germany, 2012.
- G. J. Grabovskij, T. Peichl, J. Lisenfeld, G. Weiss, and A. V. Ustinov. ''Coherent atomic tunneling tuned by strain in a Josephson phase qubit''. Poster. SOLID workshop, Grenoble, France, 2012.
- G. J. Grabovskij, T. Peichl, J. Lisenfeld, G. Weiss, and A. V. Ustinov. ''Strain Tuning of Individual Atomic Tunneling Systems Detected by a Superconducting Qubit. Workshop on Quantum Simulations, Bilbao, Spain, 2012.
- G. J. Grabovskij, J. Lisenfeld, C. Müller, J. H. Cole, G. Weiss and A. V. Ustinov. ''Two coherently coupled TLSs inside the Josephson junction of a phase qubit''. Talk. DPG, Regensburg, Germany, 2013.

Acknowledgements

First of all, I would like to thank Prof. Alexey V. Ustinov for giving me the opportunity to do a PhD in his group. I found the proposed idea very crazy and exciting and due to my already gained experience in this field I could directly start with the experiments.

I am also thankful to Prof. Georg Weiss, who extremely supported this work with his experience. He was always willing to listen and to give explanations and advices.

My special thanks go also to Dr. Jürgen Lisenfeld. He has been supervising me already during my diploma thesis and spent also a lot of time and nerves for my PhD thesis. I thank also Dr. Torben Peichl, who was a PhD student in the group of Prof. Georg Weiss and my fellow. Together we succeeded to make the experiment working.

Dr. Clemens Müller, Dr. Jared H. Cole and Prof. Alexander Shnirman deserve my gratitude for their theoretical input and fruitful discussions.

Without our mechanical (directed by Michael Meyer) and electronic workshop (Roland Jehle and Ulrich Opfer) and, of course, the helium liquefying station (Hans-Willi Pensl and Dr. Christoph Sürgers) this work would not be possible. Therefore, I thank also all team members.

My special thanks go also to all members of the Physikalisches Institut for the very nice time and very productive discussions, in particular to Philipp Jung (he took the time to read and to correct my complete PhD thesis), Susanne Butz, Sebastian Probst, Jochen Zimmer, Markus Jerger, Sebastian Skacel, Dr. Pavel Bushev, Dr. Martin Weides, Dr. Hannes Rotzinger and Dr. Oleksandr Lukashenko (our cryostats have great respect for him).

Furthermore, I would like to thank my family. Without them I would not be there where I am. My lovely Tatjana and my sweet daughter Veronika gave me energy and good mood every day. In particular, I hope that Tatjana's PhD thesis will come to an end very soon. I thank also my parents, and, in particular, my mother for visiting us and having fun with Veronika. My sister, her husband and their both sons made it possible to have holidays with full board and wellness for free and to do fitness training on their trampoline.

Experimental Condensed Matter Physics (ISSN 2191-9925)

Herausgeber
Physikalisches Institut

Prof. Dr. Hilbert von Löhneysen
Prof. Dr. Alexey Ustinov
Prof. Dr. Georg Weiß
Prof. Dr. Wulf Wulfhekel

Die Bände sind unter www.ksp.kit.edu als PDF frei verfügbar oder als Druckausgabe bestellbar.

Band 1 Alexey Feofanov
Experiments on flux qubits with pi-shifters. 2011
ISBN 978-3-86644-644-1

Band 2 Stefan Schmaus
Spintronics with individual metal-organic molecules. 2011
ISBN 978-3-86644-649-6

Band 3 Marc Müller
Elektrischer Leitwert von magnetostriktiven Dy-Nanokontakten. 2011
ISBN 978-3-86644-726-4

Band 4 Torben Peichl
Einfluss mechanischer Deformation auf atomare Tunnelsysteme – untersucht mit Josephson Phasen-Qubits. 2012
ISBN 978-3-86644-837-7

Band 5 Dominik Stöffler
Herstellung dünner metallischer Brücken durch Elektromigration und Charakterisierung mit Rastersondentechniken. 2012
ISBN 978-3-86644-843-8

Band 6 Tihomir Tomanic
Untersuchung des elektronischen Oberflächenzustands von Ag-Inseln auf supraleitendem Niob (110). 2012
ISBN 978-3-86644-898-8

Band 7 Lukas Gerhard
Magnetoelectric coupling at metal surfaces. 2013
ISBN 978-3-7315-0063-6

Experimental Condensed Matter Physics
(ISSN 2191-9925)

Band 8 Kirill Fedorov
Fluxon readout for superconducting flux qubits. 2013
ISBN 978-3-7315-0067-4

Band 9 Jochen Zimmer
Cooper pair transport in arrays of Josephson junctions. 2014
ISBN 978-3-7315-0130-5

Band 10 Oliver Berg
Elektrischer Transport durch Nanokontakte von Selten-Erd-Metallen. 2014
ISBN 978-3-7315-0209-8

Band 11 Grigorij Jur'evic Grabovskij
Investigation of coherent microscopic defects inside the tunneling barrier of a Josephson junction. 2014
ISBN 978-3-7315-0210-4